A Rock
between
Two Rivers

Fracturing a Texas Family Ranch

HUGH ASA FITZSIMONS III

Trinity University Press
SAN ANTONIO, TEXAS

Published by Trinity University Press
San Antonio, Texas 78212

Jacket design by Rebecca Lown
Book design by BookMatters, Berkeley
Author photo by Joanne Herrera

ISBN 978-1-59534-840-1 hardcover
ISBN 978-1-59534-841-8 ebook

Portions of this work appear in a different form in TribTalk.org ("A Rancher on the Border of Fear and Compasion," August 3, 2014) and in *The Politics of Hope: Grassroots Organizing, Environmental Justice, and Social Change*, edited by Jeffrey Crane and Char Miller (University of Colorado Press, forthcoming).

Trinity University Press strives to produce its books using methods and materials in an environmentally sensitive manner. We favor working with manufacturers that practice sustainable management of all natural resources, produce paper using recycled stock, and manage forests with the best possible practices for people, biodiversity, and sustainability. The press is a member of the Green Press Initiative, a nonprofit program dedicated to supporting publishers in their efforts to reduce their impacts on endangered forests, climate change, and forest-dependent communities.

Printed in Canada

The paper used in this publication meets the minimum requirements of the American National Standard for Information Sciences—Permanence of Paper for Printed Library Materials, ansi 39.48–1992.

CIP data on file at the Library of Congress

22   21   20   19   18   •   5   4   3   2   1

For Sarah

I may not know who I am, but I know where I am from.
WALLACE STEGNER

CONTENTS

PREFACE

The South Texas land we call the *despoblado*—the "wild horse desert," the land whose union with the submersible water pump blossomed into "the Wintergarden"—has a history that runs deep. This land is every bit as fragile as it is resilient. Its hardened exterior makes itself known when push comes to shove, or overgrazed bare, or plowed up for us to extract more than it was ever intended to produce.

The first white men to arrive in this country found flowing creeks and artesian springs that must have set their minds in motion. Surface water in a land of periodic drought and desiccation was surely a resource they could use to further their lofty ambitions involving land and livestock. And that's exactly what they did. In his 1812 petition to King Ferdinand of Spain, Juan Francisco Lombraño described his desired land grant as being so fertile that it was home to thousands of sheep, horses, mules, donkeys, and cows. He christened his land grant Las Isletas after the tiny islands in the nearby Rio Grande. A man's wealth, after all, was measured by the number of animals he could claim.

I wonder what Lombraño thought when he gazed at this horizon. He must have imagined an endless supply of pastureland for his animals to graze. Resting his eyes on the nearby Rio Bravo and beyond, he saw only the prospects of land, water, and wealth. He could not foresee the years of little rain, the overgrazing, and the ruin of trying to extract too much from a land that has its limits.

By the 1880s South Texas had over 3.5 million sheep, a number that decimated the available forage and ushered in the destruction of rangelands, some of which bear those scars to this day. The once-lush grasslands fought back with thorn, mesquite, and prickly pear, cultivating a land inhospitable if not downright dangerous to trespassers. One run-in with a tasajillo cactus drives the point home. Their barbs implant in your skin and cause more damage by claiming a sizable chunk of your flesh when you work up the courage to yank them out with a pair of pliers. *Leave me alone*, snarls the land.

No one imagined the consequences of the first oil exploration in the 1920s—the scale and magnitude required for this manner of extraction, as well as its reward. We've had real and imagined oil booms ever since. Some have resulted in lawsuits and bankruptcy, and others have produced vast sums of wealth for a privileged few. Regardless of the outcome, it all boils down to water and its concomitant by-products, contamination, and depletion. It's a story both ageless and potentially lethal.

In the eastern and northern quadrant of Dimmit County, where the Nueces River recharges the Carrizo sandstone, the aquifer has managed to sustain itself. Starry-eyed immigrants hailing from lands much more verdant than the county planted everything from figs to strawberries; vast citrus orchards and rows of date palms graced the burgeoning settlements of Valley Wells, Brundage, and Bermuda. The township of Catarina boasted a high school, a country club, and a railroad spur. The area was known far and wide as the Wintergarden. "Come to Dimmit County," said the booster. "It's a poor man's heaven." And it was, before the water table dropped and the expense of pumping made farming impossible.

Today this land is better known as the Maverick Basin, forming the westernmost extension of the Eagle Ford shale. Its major output of both oil and natural gas make the Eagle Ford the most rapidly developed shale play on the planet.

◆ ◆ ◆

From the highest hill on my ranch, you can see the mountains of Mexico. On a clear day the faint jagged outline of the Sierra del Burros frames the horizon to the south and west. It's a sight that evokes what mountains have always stirred in people who gaze at them from afar: the subtle suggestion of hope. At twilight when the sun drops below, their silhouette expands, and so do you. It fills you with a love of where you are.

As the sky darkens other lights appear. Orange and yellow flames, miles in the distance, dot the boundaries of the view: a half-dozen roaring exclamation points of energy unleashed, natural gas flares that burn with an intensity no simple fire could ever match, hydrocarbons the earth has compressed and held for 500 million years. They are the product of our time, the wasted refuse of our unrequited quest for more. They are the visual, physical, and emotional reminder of where this land is headed.

ONE

The First Days of Oil

It begins with my grandfather, a man I never knew. A faded black-and-white photograph of him hanging on the wall of my office at the ranch shows him sitting astride a large gray dappled horse that looks both old and steady. My grandfather is watching a blacksmith hammer a piece of iron, not to shoe the horse but to make a part for a cable-tool drilling rig. A horseman at heart, he stands at the dawn of the petrochemical-industrial age, a pioneer.

My grandfather was born in the hamlet of Thompsonville, Texas, a wide spot in the road just west of Gonzalez on old US Highway 90. He was educated in a one-room schoolhouse outside of town until the eighth grade. His father, James Sword Fitzsimons, was a dreamer and a drifter, having moved from Louisiana after the family had lost everything during the Civil War. As an amateur fiddler on the local vaudeville circuit, he was not much for providing financial security, and my grandfather had to drop out of school to help support the family when he was twelve. The next year, he saddled up and became a cowboy for hire.

He rode the Guadalupe River bottomlands and little wet weather creeks that fed that river in search of stray steers that had broken out of neighboring ranches. His enterprise was undoubtedly both unique and profitable. He and his partners, the

McGill brothers, would rope wild steers, corral them, and then contact the owner whose brand was embossed on the hide. The deal went like this: The steer had gone wild, and the owner would most likely never see his livestock again if not for my enterprising grandfather. Yet the animal was branded, clearly showing to whom it belonged. So grandfather would make the rancher an offer. He could buy back his steer for half-price or else watch as my grandfather turned the animal loose on the prairie. They all paid.

He kept this wild steer enterprise going until sometime around 1901. It was then that his partners tried to talk him into buying a ranch with them and become a proper ranchman. "No thanks," he said. He was going over to East Texas to check out this place he'd heard about called Spindletop.

At eighteen, he arrived in what was unquestionably the most violent, lawless, and economically volatile town in Texas. Earlier that year, an oil well at Spindletop, a salt dome field south of Beaumont, came in, gushing forth hundreds of thousands of barrels of oil. Millionaires were created overnight. Dirt-poor farmers became instantly rich beyond their wildest dreams, and the streets of Beaumont ran black with mud, blood, and oil, a reflection of a time when greed and lawlessness were the two main underpinnings of Jefferson County. In no time at all, Texas became the world's largest oil producer. My grandfather worked his way up the ladder from roughneck to tool pusher, and finally to owning his own rigs. The boom at Spindletop set my grandfather's course for the next thirty years. He made the transition from itinerant cowboy to oilman, extracting what the earth had been holding in its belly for 500 million years.

I can picture him going to the bank, lining up with the other drillers. They all wore muddy boots and exuded the heady scent of sweat. Over time he came to realize a longer-term and more profitable form of payment from his employers. Instead of being paid in cash, he took a portion of his pay in company stock. Luckily, he drilled for the Texas Company, which in time became Texaco, and that little bit of acumen soon turned a handsome profit for him.

Money from that oil and his early years as a drilling contractor gave him the means to buy the South Texas ranch that would, in my time, become an arena for a battle over oil and water. I never got to talk with my grandfather about it, but my suspicion is that after seeing and participating firsthand in what we now call "the industry," he was most likely determined to never let his new ranch be defiled by a driller's bit. He'd seen what they could do. The land my grandfather decided to call home was never leased for oil and gas until after he died.

◆ ◆ ◆

A few years ago, adrift, I sought out Thompsonville, the place where he was born. I drove down a twisting farm-to-market road, past the springtime green and wildflowers lining the asphalt. I found a small cemetery with tilted gravestones and plastic flowers askew in mottled vases, sunlight-cracked gifts meant to honor but not remain forever. I came for an answer, but I found next to nothing in the rock-hard ground.

I was born on the Day of the Dead, November 2, 1954. In Mexico two official days commemorate the dead: November 1 is for the children who have passed, and November 2 is for the adults.

So it is that when my birthday rolls around each year, I think of my grandfather, whose nickname was Boo.

His final tracks lead to his home in San Antonio. It was Cinco de Mayo, May 5, 1955. One of the most revered national holidays in Mexico, it commemorates the day the Mexican forces defeated the French at the first battle of Puebla in 1862, a day of victory and of reckoning. My grandfather finished his early morning coffee with my grandmother and then left the sunlit breakfast room. He strode outside alone. His right hand gripped a pearl-handled Colt .45, an engraved Single Action Army. Carved into the grip was a steer's head with two blood-red rubies for eyes. He lifted the gun to his right temple and with one squeeze of the trigger ended his life.

I didn't learn about the circumstances of my grandfather's death until I was a sophomore at boarding school. I'd been living two thousand miles away from the ranch in a world both cold and foreign. My parents had recently divorced. I took the train into New York to meet my mother, and we sat across from one another in the Polo Lounge at the Westbury Hotel. "Your grandfather Fitzsimons committed suicide," she said.

I stared in disbelief, not at her, but at the mural on the wall in front of me. Asian horsemen, crowned with brightly colored turbans and swinging mallets, charged madly across the field of play. I sat frozen. I muttered the only word I could think of. "Why?"

"Your grandfather was so disappointed in how the ranch was being handled that he took his own life."

It may have been illness or injury that drove my grandfather to suicide. It may have been depression. But I believe he found himself standing at a confluence of troubles. He had told his foreman, E. L. Pond, that he was leaving the San Pedro Ranch and he wasn't coming back until it rained. He never returned. Drought, old age, the fear of what was coming, and the knowledge that he had perhaps set something in motion that he had lost control of all must have played a part in that deadly decision.

By most accounts, 1955 was the worst year of what is known in Texas as our "drought of record." Nothing—not loss of a loved one, loss of income, or debilitating medical maladies—can compare with the anguish a rancher feels during drought. And an extended drought is a deathwatch. A deadly sentence of untold suffering to the land and animals he had tended. Boo must have felt as if God had somehow forgotten him, left him alone in a desiccated desert to bear witness to the end of creation. There was no escape but one.

What I have found from secondhand accounts, newspapers, and funeral directors, is that my grandfather took his life here just after he had breakfast with my grandmother. (The plan had been for him to be evaluated at the Menninger Clinic in Topeka, Kansas.) I can't help but return to that moment. Before my father arrived at their house to drive with my grandparents in the gigantic old Cadillac, before the luggage was set on the black-and-white marble of the foyer, before the coffee was poured and the eggs were cold. While the biscuits were still warm, and the pistol remained holstered. I didn't hear that shot, but I've felt the aftershock all my life. Maybe I could have avoided the splintering of a family and of a self that has been chipped away at its core. A lot could have been different. Maybe Boo could have been there to guide his son. Maybe he would understand his grandson's desire to return the land to the wild. He would have known well enough to guide me.

◆ ◆ ◆

Last night I invited my eighty-seven-year-old father to go to the centennial celebration of the Old Trail Drivers Association of Texas. The organization is steeped in the lore and legends of South Texas. There were stories all around, told and heard by multiple generations of landowners and ranchers whose ancestors were somehow connected to a time in Texas history that has been so completely romanticized that you might as well start attacking the defenders of the Alamo as say anything negative about a cowboy or a cow in that room at the Witte Museum.

There were stories of killings, bawdy behavior, and a lightning storm that concluded with a stampede where four different herds got tangled up and it took the trail drivers two weeks to sort them out.

My favorite bit of trail lore was overlooked. It comes from J. Marvin Hunter's 1920 book *The Trail Drivers of Texas*, which includes an interview with a cowboy who recounts his reaction to having a string of his horses stolen. The cowboy tracks the horses and the alleged thief and catches up to him just west of the Pecos River. When he rides up on the thief and sees his brand on the horses, he pulls his pistol and shoots the man dead. The interviewer is a bit taken aback by this frontier justice and asks, "You mean you didn't even take the time to ask him if he was the one who stole the horses?" The cowboy replied, "Well, no. In my experience, conversations out here on the trail can sometimes lead to trouble."

That story transports us back to a time when there was a simple code of right and wrong. If you were in possession of a horse with somebody else's brand on it, then you could be executed with impunity. Such is the romantic glow of the cowboy and the trail. All you need to know is what you see in front of you. And what I see before me from my point of view on my ranch is potentially as dangerous as an unholstered handgun.

For all the cowboy mythos, most of the guests last night are beneficiaries of the oil and gas industry. The pioneers are long gone, like my grandfather, but the industry they forged still marches on, making fortunes for some and misery for others. But then there is that mythos, the legends and lore that pass for history, a history we have in large part fabricated to fit our own ends. Oil and gas has employed thousands of Texans and doled out billions of dollars to individuals and the state; it fuels the state's finances and lines the pockets of elected officials who protect it. The industry is central, much more so than cowboying and ranching, and is the source of vast fortunes—and money is a thing that can make you turn your head and look the other way.

◆ ◆ ◆

When I was growing up, the San Pedro Ranch was a quiet place, its silence interrupted only by the singing of birds, the lowing of cattle, the neighing of horses. It is noisier now. The jarring clang of drill pipe breaks my reverie, threatening to darken my sky. But then a voice pulls me back, my two-year-old grandson, who proclaims in a voice that is both tiny and loud, "Giddy up Lombraño, I'm tauboy Leo, and you're tauboy Da."

I am here for Leo. I want my ranch to be here for him as well.

TWO

The Land

The part of South Texas where the San Pedro Ranch lies took a long time to become a place anyone would call home. For eons this land had just existed. When nomadic people came, they stayed as long as it rained. Then they moved on. The good times were when there were plenty of bison, mesquite beans, and whitetail deer to sustain them and allow them to expand their own tribe. That is, abundance fostered reproduction; conversely, drought inspired abject fear, and led to logical next steps that we prefer not to think about. Spanish explorers must have gotten here in a dry spell, because they called this place *el despoblado*: the unpopulated space, no-man's-land. Obviously, they had been expecting something else.

The ranch is a day and a half's walk from the Rio Grande, and it is where my grandfather began his life as a rancher.

At one point in his agricultural career, my father ventured over to the Rio Grande and the little riverside hamlet of El Indio. There he met one of the longtime residents and asked him what was the best crop to raise here. Without a moment's hesitation the man replied, "Newcomers." What keeps hope alive for the people down here, what keeps them coming back and convincing themselves that life can be rich and rewarding down here, is that no matter how bad it gets during a dry spell, it always seems to come back, in a miracle of regeneration and rebirth. It is the water that gives you that sense of permanence, the illusion that this land will sustain you and itself. If it is abused, nothing could be further from the truth. Big dreams and great plans are ultimately turned to so much dust, or dried and caked mud at the edge of what was once a lake teeming with wildlife and fish.

Eventually this fickle oasis, the San Pedro Springs in the *despoblado*, was given a name: Charco de los Cuervos, the waterhole of the ravens. In the journal he kept in the late 1600s, the French Jesuit priest Father Massenet remarks that thousands of massive coal-black Chihuahua ravens surrounded the San Pedro Springs when he arrived. They circled and incessantly cawed at the sight of the sojourner, who, as far as they were concerned, could just as well have been deposited by a flying saucer. No feathered creature survives like a raven, and they always know safety when they see it.

The Spaniards kept coming. Soon the Camino Real traversed the southern half of the area, providing passage from Saltillo, Mexico, to present-day Natchitoches, Louisiana.

Other names were also given to this land. Outlaws called it the Nueces Strip because it was bound on two sides by rivers: the Nueces and the Rio Grande. It became a good place to hide. Lawlessness was its watchword, and if a man wanted to outdistance himself from sheriff, wife, or recalcitrant in-laws, then the verdant canopy of the Nueces was his destination. If his crime required a more permanent solution, then the Rio Grande was the obvious choice. The famous sometime lawman and full-time outlaw King Fisher made this strip of land his private fiefdom, and the sign he posted over the Bermuda crossing just north of Carrizo Springs said it as plain as day: "This is King Fisher's road, take the other." But Fisher was generally regarded as a fair-minded Robin Hood of the brush country, a man who took care of the less fortunate while quickly gaining a reputation for rough justice. The very ground I walked on was saturated with stories.

In Fisher's day an unincorporated Dimmit County still had water that flowed from its springs. The brush had not encroached on the native grasses. But there were changes. The great bison herds that once roamed free were thinning out. The native peoples were pushed back, rounded up by Spanish clerics who used them for labor in their missions while attempting to convert them to Catholicism. This was the *encomienda* system—essentially labor in exchange for protection.

It was here in 1812 that Juan Francisco Lombraño submitted his formal petition to the king of Spain for the roughly 33,000 acres that would become part of what my grandfather purchased. He had occupied the land as a squatter since 1805, but now he was ready to become the owner in fee, and that required a petition to a higher power. Señor Lombraño claimed the land in the name of King Philip, performing the ceremonial tossing of soil and rocks, pulling herbs from the ground, and repeating a vow of obedience to a king who was three thousand miles away, a king known by his people as Pepe Botella, or "Pepe of the bottle," because of his affection for red wine.

This lawless, wide-open paradise was made for the nomad, not the more sedentary Spaniard bent on packing as many domesticated animals as possible onto his domain. Juan Lombraño stocked his ranch with thousands of sheep, goats, cattle, horses, and mules. Anything that consumed pasturage was encouraged to take up residence and procreate.

Then the Camino Real began to see different kinds of travelers. Over this road came the seekers, destroyers, priests, and pioneers who sought to find wealth in a world where the nomadic way of life had been the only reasonable approach to survival. It was on this lower presidio road that Santa Anna sent his ground troops to the Alamo, wisely deciding to keep them out of the road that led from Laredo. Civilians and enlisted men can be a volatile mixture in any era.

Just upriver from us, where Las Moras Creek flows into the river, the very first Anglo settlers attempted to build their community. In 1834, with more bravado than good sense, a small group of pioneers, predominantly German, British, and American, struck out from Indianola, in the Mexican province of Coahuila y Tejas, for the village they christened Dolores, which translates "suffering." Dolores soon lived up to its name. All of this was recorded in the diary and letters of a young German adventurer by the name of Edvard Ludicus. He was an intellectual by birth, education, and training. After landing in New York he had fallen in with Dr. John Charles Beales, who had obtained several million acres from his wife, a Mexican citizen. These men were known as empresarios, the fast-talking, pie-in-the-sky developers of their day, who were obligated by Mexican law to populate and settle their grants with European and American citizens or relinquish the land to the nation. What Beales did not take into account was that while he might have titular possession and a sheet of paper that reflected his ownership of the land, he did not control it in any way. His primary misstep was in locating Dolores a scant quarter-mile from the notorious Comanche Trace, the historic raiding trail that began in Texas and led south into the interior of Mexico.

In a desperate effort at self-defense, the colonists hollowed out a live oak tree and fashioned it into a makeshift howitzer. It shattered on the first shot, sending shards of the tree in every direction. So in less than a year, the colony disbanded. Indians, the Texas revolution, and an order from Santa Anna to exterminate all settlers who would not swear an oath to Mexico and convert to Catholicism spelled the end of Dolores.

A small group of refugees consisting of two families, the Horns and the Harrises, left Dolores and attempted to travel overland to Matamoros and escape. Mrs. Harris's account, written later from St. Louis, doesn't offer a reason for why they changed course and turned north. There they crossed the Nueces and were captured by a Comanche raiding party. Sarah Anne Horn watched as her husband was clubbed to death before her eyes. She never saw either of her two sons again, an irrevocable separation that haunted the poor woman until the day she died. Mr. Harris was killed, but not before he witnessed a Comanche brave on horseback wrest their infant son from Mrs. Harris. Grabbing the squalling child by the ankles, he dispatched it with one swift downward thrust, crushing the poor child's head on a rock. Such were the customs of the Lords of the Plains. If you were old enough to be of service, your life might be spared, but if you were too old or too young to be of value, your number was up.

◆ ◆ ◆

Just under the surface of the ranch lie the layers of the relics from those who made their home here. Several years ago Antonio Gallegos, who has been working at the ranch since the early 1970s, gave me a silver American quarter-dollar he found at the springs. It bears the date 1845, the final year of the Republic of Texas.

Runaway slaves and freedmen who were hoping to establish themselves as citizens were some of the first to try their hand at Dimmit County. In 1850 a freedman by the name of John Townsand became the first to attempt colonization in Dimmit County. The effort was short-lived. Constant harassment by bounty hunters in search of runaway slaves en route to Mexico prompted Townsand to pull up stakes and move closer to Eagle Pass and the safety of Mexico, a far cry from his point of origin in Minnesota.

The history and the saga of a particular place on this earth can be so entwined. You don't know how or when an innocent or unintended encounter with someone can elicit the most remarkable story, such as one I turned up on a recent trip into Carrizo, where my son Patrick and I went to have breakfast at a Mexican restaurant.

After soaking up the last of our huevos rancheros with Brenda Barejas's homemade flour tortillas, I asked her about the faded black-and-white photos on the wall. She pointed to one photo of a tall, proud-looking Mexican man wearing a clean white shirt and a stiff wide-brimmed hat. "That is my husband's grandfather, and next to him is Bob Lemmons, a man who was born a slave around 1850. My husband's great-great-grandfather was Bob Lemmons, and I am a direct descendent of Levi English, who was my great-great-grandfather. In 1838 Levi's new father-in-law, Mr. Burleson, gave Bob Lemmon's mother Cecilia to his daughter

Matilda as a wedding gift to her and Levi." It took me a while to sort it out, but what Brenda was telling me was that she had married a man whose great-great-grandmother was a wedding gift to her great-great-grandmother.

Several historians—J. Frank Dobie and others—have documented Bob Lemmons's life. Horatio Alger has nothing on Bob Lemmons. After gaining his freedom in 1865, Bob began hunting wild cattle and selling the tallow. Next he moved into mustanging. Riding out through the wide-open prairie and then working his way down the bottomlands of the Nueces, he would camp out for days on end, slowly and calmly integrating himself into his unsuspecting quarry. The wild horses grew to accept him and Warrior, the horse he rode to lead them, and before they knew it they had been led into a makeshift brush corral. Supposedly what won the confidence of the herd was the smell of Bob and his horse, and Bob never bathed while on these expeditions: he knew that smell said who you are, and he was the leader.

He was careful with his money. In the 1870 census of Dimmit County, Bob Lemmons had assets of $1,140, more than most of the men in the county. He expanded his landholdings and cattle herds, leasing land for grazing, and during the Great Depression he loaned significant amounts of money to Anglo ranchers in the county. He was a charter member of the Baptist church and was both humble and proud of his heritage. He died on Christmas Eve 1949 at over ninety years old.

Levi and Matilda Jane English are generally credited with being the first white settlers in Dimmit County. Levi's mother died when he was ten, and his father soon remarried. But Levi must have been at odds with his new stepmother, and she with him, because he left home and somehow managed to take up with the Comanche Indians. He spent six months with the tribe and wandered what was then Coahuila y Tejas before arriving near Carrizo in 1833. The sight of ten flowing springs gushing clear water to the surface greeted Levi after he crossed the Nueces, transforming a roving wayward lad of sixteen into an ambitious rancher who wanted to someday make this land his home.

◆ ◆ ◆

I have yet to meet the rancher who turns down mineral money, including myself. And like-minded ranchers who know that there could be some painful indigestion from the free lunch that fracking has provided are as scarce as hen's teeth. Today in many places the live oak–lined Leona River has run dry. The water table has been pumped down by center-pivot sprinkler systems to irrigate cotton, cabbage, and corn. But unlike the water used for fracking, that irrigation water stays in the hydrologic cycle and is used to produce something that creates longer-term jobs for the county.

A July 4, 1865, celebration of independence and community fellowship on the banks of the then-flowing Leona River, feasting on ham and venison under the oaks with loved ones, vanished in the blink of an eye when the Kickapoo attacked one day. During the skirmish Levi English's son Bud was killed, and his own throat was pierced by a broadhead arrow that reduced his voice to a whisper for the rest of his life. After the attack Levi gathered over four hundred settlers and led them south, crossing the Nueces and founding the village they called Carrizo Springs.

When I was a teenager, my father pointed out a small rounded stump in the West River pasture of the San Pedro, then another and another every hundred yards or so, all lined up in a row. Displaced shapes alien to the brush and scrub mesquite that surrounded them, they were the jagged and broken remains of one of the first long-distance telegraph lines in the state of Texas. The US Army built them in 1876 to connect Fort Duncan in Eagle Pass downriver to Fort McIntosh in Laredo. The man in charge was Adolphus W. Greely, a hardworking and scientifically minded individual who served in the Union forces in the Civil War and later helped organize the National Geographic Society, and who, unusually, was awarded the Medal of Honor for a lifetime of distinguished military service—despite the fact that he was the commander of an ill-fated Arctic expedition that resulted in the firing squad execution of one of his men for stealing food. The poles of cedar, pine, and juniper that Greely's men put up were twenty feet tall. They had been imported from as far away as Virginia at an average cost of $2.80 apiece. The wire was imported from England, and ceramic insulators were strewn over the ground like broken eggshells.

Once the "singing wire" had been installed, it took a lot of effort to keep it intact. The Comanches and Lipan Apaches made short work and great sport out of setting the poles on fire and burning the pitch pine that had been hauled from distant states. But the first long-distance rapid communication had arrived, and the vise was tightening on their way of life, which would soon vanish.

By 1885 Carrizo Springs, about twenty-five miles from the San Pedro Ranch and the seat of the newly created Dimmit County, boasted both a courthouse designed by famed architect Alfred Giles and a public school building that was shared between the children and the local chapter of the Masons. The booming town was ushering in an era of big ranching and small, irrigated family farms. As long as the water kept flowing, there was nothing to worry about.

But it is the original inhabitants of Dimmit County, the Coahuiltecans and their way of life, that my deepest attention has always been drawn to. On the south bank of San Pedro Creek is an outcropping of Carrizo sandstone that has hardened to the surrounding soil, yet it is still impressionable to the touch of an incising tool such as bone or antler. The first inhabitants used these tools to make their marks. In the center of the sandstone I can still see the deep hole that was used as a grinding pit for the mesquite beans of summer and the persimmons of the fall.

San Pedro Spring was certainly the center of their world. Here they knapped projectile points and wove the sturdy sotol baskets that were so essential to a nomadic life in a harsh world. Survival meant being able to gather what you could when you could, store it as best as possible, and then carry it with you when you left.

The Coahuiltecans were survivors. They understood how to live in this arid place, and they thrived in this land until the Spaniards arrived on horseback.

Recently I drove past the rocky, eroded hills that are now a part of my brother and sister's ranch. I still carry the power of those hills within me. They are replete with reminders of my past encounters with a wildness that has gone. Covered by dense stands of guajillo and blackbrush, these thickets are impenetrable to anything on two legs; they are passable only to the low profiles of the javelina, coyote, mountain lion, and bobcat. These hills are something of a geological anomaly. Smooth, polished river stones appear along with the small remnants of petrified wood that lie along what once was the ancient streambed of the Rio Grande. The wood is petrified palm, dark brown and fractured with a distinctive black and gray grain where every cell is segmented one from another. You can see how the bark of a palm tree was formed. These petrified palm remnants are elongated sections different from the surrounding rocks, with a heft and character all their own. Picking one up in your hand, you marvel at the age, beauty, and undeniable power of these once-living things, tokens of the immortality the earth can bestow.

Here nature managed to hold on to her harsh intrinsic beauty with a shield of heat and aridity that kept destruction and development at bay. There was an unadorned, enchanting depth to these rolling hills because there was nothing here that anybody particularly wanted. For a while, at least.

THREE

A Poor Man's Heaven

In 1877 a young man named Asher Richardson came to Dimmit County. He had just been mustered out of the US Cavalry, relinquishing his position as a saddler, and was honorably discharged by his commanding officer with the following comment: "This man is entirely trustworthy and of excellent character." He was born in Maryland, the son of a prosperous family with roots that extended back to 1680. But rather than return east, he settled on what he saw as a land of limitless opportunity. In 1879 he went to work for William Votaw, who owned the San Pedro Ranch. Together they stocked the ranch with more than 26,000 sheep, almost one animal per acre—a huge number by today's standards, but one in keeping with the ranching practices of the time. In the 1880 census of Dimmit County his occupation was listed as "wool grower." In June 1881 Richardson enhanced his holdings even more by marrying Mary Isabelle Votaw, the boss's daughter. This union elevated him to partnership status, and the rush to succeed and expand was on. By 1905 he controlled almost a quarter of a million acres of land, and he was just getting started.

Back in the early days, before oil, Dimmit County had a tagline: "Come to Dimmit County, Texas. It's a Poor Man's Heaven."

Artesian wells and the prospect of bumper crops fueled the hopes of clueless newcomers. The first goal of the booster was to remove the fear and uncertainty from rural life, and water did exactly that. In a 1909 edition of the *Carrizo Springs Javelin*, the editor quipped, "The man with the hoe has appeared on the horizon." That same year, Richardson made plans to build a short-haul spur rail line that was designed to bring to market everything the man with the hoe raised, completing the circle of commerce by giving the farmer access to the urban world. The line was to run from just south of Dilly to Asherton, the town he had established and named for himself. He commissioned famed Texas architect Alfred Giles to build him a Prairie School–style mansion out of native sandstone. Richardson was afraid of fire, so every bedroom in the home had two doors. The house even featured a wine cellar, a rare thing in those days. He named his home Belle-Asher and presented it to his wife. It still stands today.

Not too many years ago I walked through the side gate of Belle-Asher and visited this radiant remnant of Texas boosterism. A hundred-plus years have not diminished it in the least. Set on a slight rise just off Highway 83 with a state historical marker planted at the entrance, Asher Richardson's sandstone jewel still presides, as imposing as it is magisterial.

Flanking the sidewalk on the east side of the home, all but hidden by surrounding oaks and bull mesquites you can't get your arms around, are planted two Mediterranean olive trees, placed there by either Isabelle or Asher—or so I guess. No one had ever dared to attempt such a horticultural chimera as an olive tree in Dimmit County, and no one could have known that the tree of Athena would one day take root and become an industry. Today more than 200,000 olive trees have been planted in South Texas. They have a penchant for survival that is made for this climate.

My good friend and fellow Dimmit County rancher Jim Marmion planted several thousand trees, because as a native son he knows the ups and downs of moisture and life on the edge. That orchard stands as a testament to what one man can do. Last October we dug and transplanted five of Jim's mature trees to our new home on the ranch. Their presence instantly transformed the space, as if Merlin himself had waved a magic wand over the yard. You could see the power of those trees in the faces of every carpenter and mason who was laboring on the home.

To finance the railroad and his Asherton Land and Irrigation Company, Richardson mortgaged most of his land, including the San Pedro. He later reflected on the folly of his venture: "If you want to lose money fast, build a railroad." That's just what happened. His dreams were cut short when the city fathers of Carrizo secured a rail line of their own, establishing that town as the county seat once and for all, and leaving Asherton out in the cold.

Still, Dimmit County prospered. Newly discovered artesian water wells brought life to the parched soil. Finally, the elixir of agriculture had been tapped, just a few hundred feet below where they stood. All a man had to do was plant and water, then sit back and watch his fortunes take root and ripen.

On the front page of the July 5, 1916, issue of the *Carrizo Springs Javelin*, there is a photograph of young girls and boys standing knee-deep in the cool, clear waters of the town springs, which once watered bison, and later longhorn, antelope, and javelina. These waters now nourished vegetable gardens and citrus trees. The Baptists baptized here, and the spooners spooned on warm summer nights as the songs of cicadas filled the air. The *despoblado* was turning into Texas.

Real-estate sharpies, those great avatars of boosterism, tied grapefruit and oranges to the mesquite trees that lined the tracks of the railroad. Speeding past, unsuspecting Yankees would marvel at the fecundity of a land whose nature was to produce no matter what the circumstances. The boom was on, and the fuel was water, seemingly endless stores of it.

In 1926, in the tiny town of Catarina just south of Carrizo, the greatest land promoter in Dimmit County history made his mark. Charles Ladd was a tall and dashing salesman who wore cavalry jodhpurs and knee-high stovepipe cowboy boots. Crowning his regalia was a ten-gallon hat creased to perfection. In a photograph from that year three young women stand next to a stately Moorish-style home, built by Ladd as an example of how perfect life could be in Catarina. But these ladies must have been more than a little perceptive for their age, because the caption under the photo reads, "The house where suckers are hooked."

With a telephone exchange, waterworks, a stucco and Spanish tile high school, and even the Catarina Country Club, Ladd did his best to entice buyers into his dream. At the height of his exuberance he constructed an imposing entrance to the orange and grapefruit orchard he had planted. The inscription on the arch says it all: "A-Ladd-In Garden." But no matter how hard or fast his followers rubbed the lamp, they could not revive the groundwater that was dropping every day, and Catarina all but disappeared. What remains is a convenience store on Highway 83 that serves the oilfield.

As I sit here, the rumble of eighteen-wheelers seeps in. They are hauling caliche to a new pad site where the extraction business continues. The first time anyone drilled a well down here was in 1924. It was a bust, and the driller ran off with the investor's money. It was an old cable-tool rig like my grandfather's, fired by low-grade lignite coal that was mined just across my fence line on the Dentonio Colony. By then Dentonio was already crumbling, its water played out during the drought of 1918–19. That should have been a warning.

In 1914 Asher Richardson died, leaving a debt that would hang over his children until the Frost Bank called the note and my grandfather bought the ranch out of bankruptcy in 1932. The booster had had his run, and as long as the water flowed there would be no shortage of prospects who believed in the power and limitless possibilities of a land that had been dubbed "the Wintergarden."

With every new well the pressure in the groundwater was reduced, and what had been artesian quickly became less than economical. As the cost of pumping proved prohibitive, more farms succumbed to dry weather and the railroad's high freight charges. Nothing satisfies a man like the illusion that because he has a little water now, he can take on whatever curve balls nature throws his way. The cold hard fact is that you can't pump your way out of a problem. In the prophetic words of Benjamin Franklin, "You shall know the value of water when the well runs dry."

◆ ◆ ◆

The San Pedro Ranch took its name from the desert's greatest miracle: an artesian spring. Whatever geological event had ruptured the surface of the land and released the hidden water, it changed things. The spring found its way up and out, and water flowed.

From San Pedro Creek's hidden source, the fragile stream flows south and west. In wet years it always collects in a pool of limpid, pale green water a hundred yards long. At the far end of the pool is the actual crossing of the Camino Real. As was their custom, the Spanish named springs for the feast day of the saint that coincided with the springs' discovery. Ours was named for Saint Peter, the patron saint of fishermen.

Below the pool rests a perfectly rectangular large stone that serves to contain the waters of the creek. I have an affection for these old weir dams. They contain just enough water to serve mankind and animals, allowing plenty of overflow to pass downstream and keep the creek healthy.

When my grandfather bought this ranch, there was a single windmill in the North Mott pasture casting a silhouette across the bluestem prairie. That wooden tower remains here today, a touchstone to a time when sixty-seven feet down was all you had to drill to supply your family and your livestock with the water they needed to thrive. Off to the side of the mill, resting as if it had emerged from the surrounding Carrizo sandstone, stands a worn-out remnant of that time. It is a solid rock *pila* with a schemer of chipped cement holding the rough rocks in place. They are as solid as they are timeless and beautiful. Tar from another time makes black tracks down the interior where water once graced this vessel of survival. Inside the structure, cactus and blackbrush have taken the space and made it their own.

The scene reminds me of a diorama one might encounter in a natural history museum. The windmill is named the San Francisco, for the patron saint of all animals, and was christened by the Votaw family, whose ancestors had owned the original Spanish land grant.

For decades it pumped a meager two gallons a minute to fill the pila for livestock and wildlife. At the base of the San Francisco, where the casing of the windmill joins a small pad of evenly poured but now crumbling cement, is etched a tiny reverse swastika that the driller or one of his workers made in the newly poured wet cement. Native cultures around the world have made this symbol theirs, and the meaning is universal: "May this place have good fortune and fair winds."

The well gave just enough water to sustain animals within a four-mile radius. In 1927, when it was drilled—five years before my grandfather bought the ranch—the San Francisco must have been considered a bold step for a rancher to take. By using this relatively new technology to increase the carrying capacity of the land, he was able to double the size of his herd. By most accounts this was a modest and relatively low-impact use of one's own resource. But humans have never been known to be self-limiting creatures.

By the early 1920s, the cost of pumping marginal water wells was reducing the number of cultivated acres in the county. Shortly thereafter, the cow pastures that had been converted to row crops were once again being stocked with cattle. Around 1923 the first serious exploration for oil and gas in the county began. A partnership out of Fort Worth named Wheatly and O'Hearn leased the San Pedro from Asher Richardson's widow Isabelle and spudded a wildcat exploration well in the southeast corner of the Lombraño grant. The widow was in danger of losing the ranch, and she was banking on saving her family's land by producing what was beneath.

Those Carrizo sands may have been depleted of water, but the dream of hydrocarbons was beginning to take hold. In the July 20 issue of the *Javelin*, the newspaper's editor noted, "The purr of the wildcat can now be heard in our county."

Every week for seven straight months the Wheatly and O'Hearn wildcat was front-page news. From September 23, 1923, until mid-February of the following year, the newspaper acted as a virtual marketing arm for the two men. The partnership utilized the relatively new technology of the rotary drill. At one point, they outfitted the rig with a diamond bit to help cut through the cap rock. They employed a Dutch geologist, Bertrand de Graafe, who gave wildly optimistic pronouncements as to the rich deposits they would surely encounter if only they drilled a little deeper. In time, though, disappointed by the lack of results, Wheatly had had enough and wanted to stop drilling. O'Hearn disagreed, and the bickering progressed into a lawsuit that left the company insolvent, which left Isabelle Richardson holding their ticking time bomb of a mortgage note to the Frost Bank until that day my grandfather walked in with cash in hand.

Up to the 1940s the oil activity in the county was relatively quiet. My grandfather did not lease the ranch to drilling companies, and the few surrounding wildcatters were subdued by the reality of either dry holes or meager production, wells that came in like a house afire but dwindled to less than a few barrels a day in a very short time. But in 1943 several wells came in at relatively high pressures and production rates, which naturally sent land men to the courthouse to research mineral deeds. At the time, the talk at the Blue Quail Cafe centered on who was making leases and how much they were being paid in bonuses and royalty. Visions of shiny new pickups and enhancing one's breeding operation with some high-powered purebred bulls crept into the waking thoughts of the men at the table. But unless you were drunk or full of yourself, you would not reveal the price you got for a lease. Such talk could either be interpreted as chest thumping or, in the event he accepted too little for the lease, get a person branded as a fool who had been hoodwinked.

Old ranchers with sweat stains on their short-brimmed Stetsons would hover over cracked white porcelain coffee mugs and ponder this news. They would tilt back their hats and let go with a low soft whistle as they did the math on what they regarded as free money. The news held out hope, a safety net of sorts that would extend the illusion of agricultural independence by giving a landowner the surety of a financial foundation. "Finally," they would say out loud, "we've found a way to make this country pay. We can make it through the hard times of a drought and then restock the pasture when the rain comes back." Payments to landowners came to be known as mailbox money. All you had to do was sign on the dotted line, then sit back and wait for the postman. What stuck in people's minds was the initial production of fifty or sixty barrels a day and the belief that their well would be different than all the others—that the prosperity generated by such geologic good fortune was going to perpetuate itself indefinitely.

◆ ◆ ◆

It has taken nature millions of years to produce these resources, oil and water both, lying miles underground, sealed off by capstones of shale and rock. Men are coming for them now as never before, bringing machinery and money to a land that has seen all of this in one form or another many times before, but never like this. It is unprecedented and transforms my world. The men come, bringing leases that give them the right to scrape, alter, drill, and disrupt. Then they put their knowledge of human nature to work. A big check, "Pay to the Order of" followed by the name of a ranch in bold type, and the deal is done. Pipelines, compressors, trucks by the hundreds.

Daily I pass by the remnants of which I write. On the far southeastern corner of the ranch are the remains of the Dentonio Colony, the largest and most ambitious real estate development in Dimmit County and the first to experience complete failure. And nothing instructs like failure.

Graham Denton, a dashing and completely delusional land developer, was born in London in 1862. He came to Dimmit County, like so many others, to make his fortune. Others hitched their dreams to Denton's, and tracts of land ranging from 80 to 160 acres went to them for a couple of hundred dollars or even less. There was an added incentive: if you bought more, you received a free lot in town. Denton sold land all over the country, but by 1912 only seventeen families resided here.

One was the sister of the most celebrated Indian captive in Texas, a Mrs. Miller. She recounted the life of her sister, Cynthia Ann Parker, and her bout with a return to civilization that was both unwanted and debilitating for her. Cynthia Ann spent the rest of her life sitting under a live oak tree, not uttering a word, gazing into the distance as if looking for her adopted people to come and take her back. She was just one of many broken people in a land fertilized with hope and hard work, slapped back to reality by what has always been here, a five-year cycle of wet to dry. Feast and famine, chicken one day and feathers the next.

My father took me out in the pasture one day, many years ago, and showed me the abandoned spot where the first oil field folly had taken place, where drillers searched for what they knew was there in abundance but would not give itself up. The site was a sad little hole in the ground with a rusty pipe protruding in vain from the hard-packed sand around it.

· FOUR ·

Rio Bravo

Eight and a half miles from the border of my ranch lies the Rio Grande, the big river. I prefer the name the Mexicans give it, Rio Bravo, which means something like "river with spirit." I have always looked to it for solace and direction. There is something in the nature of its flow that makes me feel as if the water is, by its nature, a godlike force that not only soothes my mind but also heals my heart. I look to it like some people look to the Bible for relief from worry, and it has always given me benchmarks and milestones to measure myself against.

The river flows wide and deep on certain stretches and constricts to shallow fords that are the passageway for travelers north and the bane of the Border Patrol. It's a force to be reckoned with. Water that roils with the sudden downbursts of summer deluges unleashes tons of sediment from the banks and turns the water the color of chocolate milk. Branches, cane, and inner tubes used for immigrant crossings are swept toward the gulf as the river scours and cleans its channel. In time, the sediment settles and the green appears again. During the spring and summer months the river is home to snapping turtles the size of hubcaps and channel catfish that drift the placid currents feeding on the less fortunate. Forty-pound-plus yellow cats are not uncommon, with gapingly long horizontal slate-gray lips that can languidly vacuum just about any living organism within the fish's domain. On the banks of the river, wide, impenetrable swaths of carrizo cane shelter deer, javelina, and the occasional mountain lion. I once witnessed a nine-banded armadillo that I swear was the size of a half-grown hog waddle lazily through a stand of cane upriver and out of sight.

The Rio Grande has been dry from El Paso to Ojinaga since 1949. For more than three hundred miles it is a muddy ditch that is given life again only by the blessing of our sister river, the Rio Conchos, which flows north out of Mexico and meets the Rio Grande near Presidio. The Conchos is a turgid torrent of water that once fueled fields aplenty in both countries. With its new-found transfusion the Rio Grande resumes its steady roll onward through Santa Elena Canyon in the Big Bend on to Boca Chica and the unbroken horizon of the Gulf of Mexico.

Fishing with one's father stays with a man. When I was six years old, I watched my father fish for brown trout near Creede, Colorado, at the headwaters of the Rio Grande, under craggy cliffs and mountains that sheltered a verdant river valley. I sat under a ponderosa pine, perched on a rise high above the river, and watched the river below, how it coursed between boulders and rapids, the white foam of moving water rushing with a strong purposefulness that could not be contained. Below me I could see my father, a speck of a man flailing away with a fly rod at what he could not see. What I remember was how small he seemed, far below. That image of him has always stayed with me: a small man methodically moving an arm that moved a swaying looping line angling for the unforeseen and dwarfed by a great river.

◆ ◆ ◆

On the bank of the San Ambrosia Creek, I kneel in the dry grass and pull up close to the modest shade of a small mesquite. Opening my small plastic mottled green tackle box with the gold clasp, with the pungent fumes of last year's catfish bait and pork rinds wafting around me, I assess my arsenal of lures. Decisions are made on names rather than utility: a "Hula Popper" or "Lucky Thirteen" appeal to a young boy even more than the fish he hopes to catch.

We are bound by the San Ambrosia Creek, a wet-weather tributary of the Rio Grande, on the west, and the San Lorenzo on our east. Saint Ambrose is the patron saint of beekeepers, beggars, and learning. In 1691, when Father Massenet first laid eyes on this tepid backwater of the Rio Bravo, he believed that it portended a soft and gentle journey up the Camino Real. But it did not.

In a dry year the creek is but a muddy remnant of its former glory. It holds an exalted position down here, where water is as precious and as tenuous as life itself. On either side of the creek are giant native live oaks. You can see them for miles, for they stand out with color and height that set them apart from mesquite and huisache. The first Spaniards here planted oak acorns so future travelers could see the trees from that distance and find water. In the fall deer and hogs frequent the groves, vacuuming up the precious acorns that litter the ground beneath the thick dark green canopy.

Wild Rio Grande turkeys found this spot long ago. It was one of the few places on the ranch that provided the height they need for roosting and staying out of the reach of coyotes and bobcats. As the creek waters move south toward the Rio Grande, it becomes wider and longer, interrupted at intervals where sandstone rises, creating a sculpted natural dam. Willows drape and rustle in the breeze. It was here that my friend Jimmy Rutledge discovered and identified the westernmost native stand of eastern gamagrass, a forage grass so dense in nutrients and protein that it was virtually grazed into extinction by cattle. Here it hid away, in the perfect habitat to do so, sequestered by periodic flooding and drought. Its roots can penetrate a dozen feet and more of soil, anchoring a plant that once covered vast Texas prairies, a grass whose sweet wide stems fed millions of buffalo.

◆ ◆ ◆

In 1999 my father sold the West Mott pasture. It had not only the San Ambrosia but also the famed Lost Lake, a natural deposit of Montmorillonite clay that encompassed well over an acre of shallow water and mud, an elusive and mysterious body of water that was a favorite lair for mountain lions and wild hogs. The day of the sale was the last time I was ever in that pasture. I don't know if I can ever stand to go back again.

Before the sale, I made one last desperate plundering foray through my adolescent playpen. Starting at the water gap at the East River, I walked north along the creek. On the ground before me lay the chipped and shattered remnants of another people's time on the San Ambrosia. Broken arrow and spear points were everywhere. Scrapers, awls, and the fragments of drills with the worn indentations of where palms and sweaty hands gripped and worked hide and bone were spread before me for the taking. I gathered every remnant my pockets would hold, wanting to secure them—their past, my past—for the ages. Like so many others out here, after all, the sale was to a city-dwelling lawyer who wanted to have this land just long enough to sell it to someone who wanted it more than he did.

Off a high bank overlook, where the creek straightens out and the live oaks recede, is a small plateau that extends for perhaps fifty feet square. The rock is different here. Rounded river stones lie here and there, smoother and harder than the native stone. Under a small ledge of sandstone there I found a pestle, unmistakable in its form, its end abraded by years of pounding mesquite beans and nuts into coarse flour. It was beautiful, dark, heavy. The grip fit me perfectly, conforming to fingers that it had not felt for a thousand years.

The San Ambrosia has always been a natural passageway into the interior of Texas. Contraband has always been its currency, and the shelter of dense stands of brush and thickets that could stop a train have protected and concealed generations of smugglers. Smuggling down here is as natural as breathing. This river is a sieve and honeycombed with trails and paths that have led from supply to demand for centuries. Deep, tall contours of cane line the river on the American side. Here people stash thick-walled plastic garbage bags of pot, or the ubiquitous truck-tire inner tube that inflates to float its passengers across. The canebrakes are littered with them, abandoned detritus and silent reminders to just how futile the task at hand has always been.

Not long ago I was walking the bank of the San Ambrosia and found the scattered remains of one man's venture. In the early 1920s the Texas Rangers made a name for themselves as enforcers of the Volstead Act, and somewhere out there, our foreman told me, was a lost wagon train carrying tequila. He was right: we found the frame of a wagon, along with pieces of the thick leaded glass that was once filled with amber fire. As an occupation it surely must have been tempting, for smuggling tequila during Prohibition could net a cowboy in one night what would take him one month to make working cattle.

On the desk before me I have a piece of that beautiful old glass, a pale light greenish shade that holds light rather than letting it pass, thick where it has cracked, exposing a strong condensed molecular structure that was made for transporting what was both beautiful and forbidden.

◆ ◆ ◆

A fisherman's approach to life can be discerned from his tackle box: the orderly approach to obtaining what he cannot see but knows is perhaps lurking just below the surface, back in the canebrake or around the bend where cattails arc and bend as the breeze moves.

In my box is a magnificent deep royal red Garcia Ambassador bait-casting reel, a device as likely to implode and ruin a day as was ever formulated by the human hand. The trick is to adjust the force of one's cast and the flight of the lure to the resistance of the reel as it gives up its line.

Now, on this warm spring day, a bass flips the surface under the overhanging willow branch. I curve a perfect arcing cast that lands my Hula Popper just so. I let it sit until the remnants of the ripples absorb themselves back into the green water and give it a deep gurgling pop. Nothing. A fat grasshopper clicks and rattles like some airborne jalopy, landing on the limb beside me.

A rise on the opposite bank signals a new opportunity. I retrieve the Ambassador, loosen the stainless tension knob to get more distance, and let it rip. The popper arcs through the air with ease, then dies in midflight. I know what I will see before I even look. My beautiful red reel has given birth to a tangled mass of monofilament, the classic bird's nest. I let out a sigh of disgust. The grasshopper vaults for the far side of the creek and lands short, lost forever as he disappears below.

I retreat to Daddy, hand him my rod and reel. He takes it from me and smiles, pushing back his flat-brimmed straw hat and adjusting his glasses. Without a word, without a glance in my direction, he begins the task of a parent, the slow, deliberate study of a tangled mass of passion gone awry. Where good intentions meet with too much mustard and the only response of the responsible is to go about the business one is meant to do. Untangling the past, to let me cast again and give rise to the hope and quiet satisfaction of landing what remains below the surface.

◆ ◆ ◆

The river first tested me in an all-male fishing trip orchestrated by my father when I was nine years old. Our fishing companions were a wiry old World War I vet, Henry Catto, and two brothers, Jim and E. L. Pond. Jim was the county game warden, an excellent choice given our purpose and position on the river, and EL, the manager of my father's ranch, had been with us since my grandfather's time.

We were in a landscape as wild as I had ever seen—a bold rushing river with cut banks and islands of willow, tall bunch grasses, and thick cane on the bank. Here the wildlife seemed to be wilder, and the river moved with a swiftness and force I could clearly see. Limbs of trees sailed by, and coots and puddle ducks struggled against the current to find a feeding hole. Satiated, they would float lazily back downstream before doing it all over again.

Here, on the river with my father, I found a time outside of time. It was summer, and regimented schoolwork and the shame of academic underperformance that characterized my school years was mollified and almost vanquished by clear air, water, and the strength of fast-moving water. I was as wild and free in those few days as the river that flowed past us.

As the heat of the day came on as only a South Texas summer can, I retired to the campsite and started popping the tops on the bottles of Sprite the men had packed in the cooler. I could not believe my good fortune: I was miles from my mother and I had all of the sugary soda my heart desired. I drifted off to sleep with visions of yard-long yellow cats tugging on my line.

I woke up to an excited shout. "Sunny, Sunny, wake up, grab your rod, you got a whopper!" I ran to the edge of the river and grabbed my rod just as it was being pulled from the bank into the current. The cork was dancing furiously, first down, and then to the side, then down again and under. There was a real fish on the other end of the line.

As the men gathered round slapping my back and shouting orders, the fish emerged. Breaking the surface of the river, his long back parting the waters, came a yellow cat that was every bit of fifteen pounds. He was half my height. With every crank and pull my smile grew wider and my heart raced until I fell back on the bank pulling the cat to the shore. With a heave-ho, EL helped to hoist the angry leviathan from his home into the light of day.

He hung the fish from a mesquite tree so I could pose with

my prize, but just as the shutter snapped I touched the fin. The fish responded with a final jaw-splitting slap that sent me to the ground and permanently reshaped the brim of my new straw cowboy hat.

It turns out that while I was dozing in the shade on my cot, EL had slipped over to the trotline and unhooked a fish. He then reattached the cat to my hook and started yelling that my cork was going under. He had gotten me, hook, line, and sinker.

I had caught, or more precisely had landed, the fish. That was a feat, but nowhere near the same as hooking the fish. Any fisherman worth his salt will tell you that setting the hook is the definition of catching a fish, and doing both is a must if you want to take credit. So that evening after the men sat around drinking Bohemia, I struck out on my own. I strapped on my weapon, a stag-handled hatchet-knife combination, the perfect survival tool for the freshly minted summer warrior with a bruised ego. Down along the river I cut a length of cane and began to fashion a spear. The nodes on the cane were tough, but after a few deft strokes the point was sharp enough to penetrate. The handle needed some smoothing, so I flipped the knife over and chiseled the cane, pulling the blade toward me to remove one reluctant bump.

The node gave way and the blade came at me. Stainless steel met the palm of my left hand, and the flesh parted. Blood flowed. I was too stunned to scream. Somehow I managed to sheathe my knife and run for camp.

It was a deep cut. The blood poured freely as I rushed to my father. He wrapped my hand in a T-shirt to stem the gush and then handed me off to Jim, the game warden, who was first aid–trained and battle-tested as an artilleryman in Italy. He poured on a dose of iodine, wrapped a thick layer of gauze around the wound, and tied it tight. He instructed me to hold my hand above my head. I had failed as a fisherman, and I had brought an abrupt end to the entire fishing expedition. But what I remember most was my loss of face as a man in the presence of my father.

I carry that scar on my palm to this day, a white slash across the heart line. It has faded, and the pain is gone, but the memory lingers. It has become part of the story of a father and a son but also of other, less visible markings. Like San Pedro Springs and other artesian waters, its source lies hidden beneath the ground.

◆ ◆ ◆

The river cuts a wide green swath as it flows down through cane-brakes and sandbars choked with willow and mesquite. When I was nineteen years old and had dropped out of college for the first time, I ended up down here for the night.

My girlfriend at the time was a cross between Rachel Carson and Jane Goodall. She was in touch with the land around her. Birds, flowers, and everyday brush that the rest of the world takes for granted had meaning in her life. It felt good to be around her.

Everything seemed bigger on the river. A great kiskadee fly-catcher gave us a morning call that all but shook the leaves from the trees. We watched a raft of green-winged teal glide past us in the slip of the current, feeding with ease and no real desire or hurry: an arch of the neck, a preen, settling down into a quiet float before paddling back upstream to repeat the performance.

The wildlife held us that day. Suspended in reverie, we were secret witnesses to a world that was both wild and undisturbed. I could not even kiss her. As the sun broke the horizon from the gravel hills of Old Mexico, and calls of ducks and doves surrounded us, I wanted to follow teenage instincts and the desire I know I should have had. But nature held me back that day. Her gentle dawning and the rapture of unfolding beauty around me left no room for romance. I was a part of life that day, like watching the birth of a small bird, silent, small, and surrounded by a world so big.

I can still remember a time when water seemed inexhaustible to me.

There is a photograph of myself that I first saw under the glass of my grandfather's roll-top desk in my father's office in 1958. I was three years old, dressed in little overalls and holding a small red rubber ball. In the photograph, a sun hat shades my fat cheeks, but it could not begin to hide the broad smile that spread across my face. I'm standing ramrod straight and as proud as can be in the Lower Holloway pasture. My father, the photographer, was using me as a gauge in that picture to show the height of the sorghum alum grass he had planted that spring. The cracked soil and desiccated ground of the "drought of record" vanished that year like the bad dream that it was. Cattle dotted the landscape as far as you could see. The pasture had been root-plowed, raked, and burned. Then, right on cue, it started raining and did not stop all year. Rain had returned to Dimmit County. To be exact, thirty-one inches fell during that year, the most since 1949.

Another big water memory comes from the early fall of 1964. My family was all under one roof at the time. Even my grandparents from Wichita Falls were with us at our home at the ranch house on San Pedro Creek. The soil was saturated from an earlier August rain, so when a storm slowly wormed its way out of the gulf and stalled with South Texas in her crosshairs, we were in for a meteorological event of thunderous proportions. What I remember is going to bed with the faint *pat, pat, pat* of rain on the roof. When morning came, the first thing I heard was the wind and rain working in violent rhythmic concert. Gusts of water blew against the house with gale force. Family lore has me going to the window and looking out past the wall toward the big live oak one hundred yards distant from where I stood and announcing, "The River Nile has come to the San Pedro."

Indeed, there was now a river where there had been a gentle creek. Cracked willows uprooted by a force of nature unknown for decades floated past me as I stared in astonishment and boyhood wonder at the transformation of a land I thought I knew. The storm scoured sandstone and swept away what was not anchored down. It was strange to feel this new fear of flood in a land of drought. Water gave us everything; it could also take it away. We huddled together as if we were in the path of an approaching tornado, bound together by danger.

There were two somewhat aimless yet amiable brothers who worked at the ranch—the cook's sons, Mrs. Hawkins's boys. Mr. Hawkins had told me what he thought of the boys as he scowled and squinted his beady eyes before spitting into the yard: "All those boys are good for is braiding quirts. They just sit around all day and twist leather together." Perhaps so, but they did rise to the occasion on the morning of September 17, 1964.

The grownups had started to ponder an evacuation plan. The water was still rising and the rain was still falling. It came down in seemingly endless sheets, blowing sideways in sudden bursts against the windows. At some point there was a brief lull in the storm, and we all walked down to the creek.

On the other side of the bank stood the Hawkins boys. They were trying to yell, but the roar of the creek drowned out their words. So they drove off in their truck. In ten minutes they returned, taped a message to an arrow, and drew back the bow. The arrow arced over the creek and landed a few feet from my father. The note told us to drive out the back route through Faith Ranch and on to Laredo.

By now the rain had ceased and we realized our plight was not as dire as we had thought. But it had been an episode of intense danger, and the thrill of it still lingers fifty years later. I can still see that white aluminum arrow with its red and blue fletching slice across the sky with ageless grace and speed, bringing us a message. It's what a nine-year-old remembers.

FIVE

Boyhood

A boy's rural rite of passage in Texas usually begins on a horse, if he has one, and with a gun, preferably his own. When I was in the third grade, my father gave me a Sheridan pellet gun, and I broke it in by freely firing on anything that moved. Songbirds, quail, dove, and jackrabbits were all in danger as I took aim. The lucky ones got away.

It is such a mysterious process for a boy to become a man. It takes a delicate collision of a partially formed character and the right circumstances for the transformation from amorphous adolescent to a person with a clearer notion of who he is.

Ranch foreman E. L. Pond had a lot to do with the way I grew to be a man. He was hired by my grandfather at age fifteen and had been with the family ever since. Both of my parents being only children, I had no uncles or aunts to coddle me or set me on a particular path, so I was often handed off to EL. He was six feet seven inches of love and trouble, the perfect companion and playmate for a ten-year-old spoiled country club brat like me who, when he got to the ranch, wanted nothing other than a fast horse and a full box of bullets.

EL had a little one-room house over at the headquarters, and at his place I found a kind of sanctuary where the only rules were the ones you learned the hard way. In the winter there was the potbellied stove that roared and hissed as mesquite logs burned and warmth seeped from the stovepipe into the cinder-block room. With the covers pulled up to my chin and the smell of soiled sheets that had not seen a box of Tide in a month, I was in heaven.

EL would sit on his little stoop outside his room and fire up a Pall Mall at the end of the day. Cracking his Zippo, he sent sparks to the wick. A heavenly wave of Ronson lighter fluid drifted in my direction, and a two-inch orange flame burst forth. Drawing on the cigarette, he would cup his hand and squint across the yard in the direction of the bunkhouse, his small eyes clear and blue behind that veil of smoke. I watched as the smoke followed the lines and crevices in his weather-beaten face, studying the facial map of the man I revered.

On summer nights, with the mammoth Friedrich window A/C unit droning precious cold air across my face, we used to sit and watch the one channel we could get on his ancient Zenith TV. After we wiggled the antenna just so, and crinkled the tin foil for the third time, *Louisiana Hayride* would slowly emerge on the snowy screen. My grandmother had given her old set to EL as a substitute for company because he had no wife, and as a substitute I guess watching Dolly Parton parade on stage with Porter Waggoner and the Wagonmasters could be construed as companionship.

After the show, when Porter and Dolly had retired and the Gillette shaving commercial had had its run at us, we switched off the tube and turned on the reading light. EL's cow dog, Boy, would curl up by his bed, and EL would usually disappear into some Zane Grey or Louis L'Amour western where the good guy got the girl but not the ranch. Or the cattle baron got his come-uppance when the squatter won the deed to the ranch in a poker game. For me, there was only one literary destination. *True West* magazines were just as they said. Footnotes and a bibliography were not required for the buried treasure of Cochise or the last days of Sam Bass. I would slowly lose myself in sleep, and *True West* would fall to the floor.

EL had come to the ranch quite by accident. A bad one.

In spring 1942 he had decided to enlist in the Marines and take on the Japanese. Being so tall, he had no trouble lying about his age, and with his parents' consent he went to San Antonio and signed up. As a favor to his parents he helped with the spring planting on the family farm while he waited to be shipped out. One morning EL piloted a little Ford tractor down to the plot of newly planted peanuts, and cradled in the crook of his arm was his father's old 12-gauge shotgun. He planned to defend the tender green shoots from the crows by administering some frontier justice: the plan was to kill as many as possible, leaving their broken bodies scattered on the ground as a warning to their brethren.

Distracted by his targets, he ran the tractor aground onto a planted row of the very shoots he was defending. He hit the brakes hard, and the shotgun fired, the force of the explosion knocking him off the tractor as the lead pellets struck him just above the right elbow. His tattered limb trailed blood and small chunks of bone and flesh, and by the time EL and his father reached the hospital in Crystal City, he'd lost so much blood the doctor said not to bother with surgery as he wasn't going to live. But after two transfusions from his father and one from his uncle, EL looked like he might make it after all. When EL's father agreed to sell some of his cows to pay for the operation, Dr. Poindexter relented and amputated what was left of the arm.

EL was still due to report to San Antonio, and his presence caused a stir as he stood in line with everyone else. The consensus among them was that we must have been losing the war badly if we were now accepting one-armed men into the service. They asked the recruiting sergeant, What in the hell is a one-armed man going to do for the war effort? "I'll tell you what he is going to do: he's gonna tell that blind boy back there who is filling up that bucket when his bucket is full."

Back home, as the hard reality of his condition settled in on him, depression and thoughts of suicide took over. He refused any company and ate alone. He would leave the house in the morning and spend the day with the cows or at the pigpen watching the animals. At night his mother would leave a plate of food on the front porch and he would squat there, eating with his hand, looking over his shoulder, dreading the thought of human contact. He was going feral.

The small herd of cattle that Ed Pond had accumulated over the years not only paid for EL's operation, but also played a significant role in saving his life. From sunup to sundown EL would follow the herd as if he were one of them. On foot he would go to the tank and drink water before lying down under the shade of a mesquite and nap as they chewed their cud and rested during the heat of the day. He even adopted their diet, foraging sweet purple pear apples from the cactus and the mesquite beans from the trees that ripened in late July. The human world remained at the edge of his existence. Animals were his salvation, accepting him without judgment. Unlike the humans, they let him into their fold.

It was on this ranch that EL proved himself and began to start his life over as a one-armed cowboy. It happened in a rushed moment, with Ed Pond racing to tell EL of an opportunity to work for current ranch foreman Bob Grissom. EL tried to bolt but was caught and wrestled into the truck cab by his father and two brothers. They stopped at a filling station and sandwiched EL into the front seat of Bob's truck between him and another cowboy. For the next twenty-eight miles Bob Grissom sternly applied every combination of both tough love and encouragement—if EL was going to make it in this world, he had to get going.

He relearned roping, using his left hand to make a loop while he held the coil under his stump, and within two weeks he was catching his own horse for a day's work. The cowboys were helpful for the most part, all except for one. EL first related this story to me about two years before he died. I had come by for one of my regular visits, and we were sitting in his trailer house on the outskirts of Carrizo just catching up. He turned to me and said, "Say, Sunny, did I ever tell you about the time I learned to push a wheelbarrow with just one arm?"

He was taunted into the task. When the construction foreman on the work crew said to no one in particular, "Do we have a wheelbarrow around here?" EL responded by saying he knew where one was and would be right back with it. When confronted with the reality of needing two hands to maneuver it, EL wrapped his leather belt around his stump and secured the other end to the handle of the wheelbarrow. He proudly deposited the wheelbarrow in front of them and returned to his earlier task of shoveling cement into the mixer. But one man said, "Now when I get it filled up, you take it out there and dump it." The gauntlet had been thrown down; EL could not help but do the impossible.

For the next four hours he pushed an eighty-pound load of wet cement fifty yards to the pens, where he then proceeded to pour the cement and set ten-foot cedar posts in place. By the fourth trip the leather belt was cutting into his flesh. He wrapped another piece of cloth on his stump and kept going. Two hours later a large, bulbous swelling surrounded the belt, forming a blister that then burst, soaking his khaki shirt in blood. That was not the only damage inflicted by that belt. But he made it to lunchtime, then retreated to the bunkhouse, where he changed shirts and wrapped his bloody stump in a handkerchief. Lunchtime chatter turned to EL, and Bob Grissom took the opportunity to weigh in. "Yeah, you sons-a-bitches thought you had him this time, didn't you? If I had two more like him, I'd run every one of y'all off this ranch today." EL had earned his place.

My father may have given me my first gun, but it was EL who taught me how to use it. We drove out on the ranch in his battered old GMC truck with the windows rolled down. I loved nothing more than the thrill of the hunt. I set my sights on deer, javelina, or wild hog. Being a hunter was how I defined myself at the time. In many respects I was no different than any other boy my age. Since my formative fishing trip had ended in humiliation with a deep cut on my palm from mishandling a knife, I would have to prove myself by hunting game.

Early one morning on the way to work cattle at the Holloway Pens, EL introduced me to the Mexican eagle, the caracara. We were driving along a fence line in the Horse Trap, and he stopped the truck. He leaned out the window, pointing his stub in the direction of a bird that was as alien as it was exotic. I stared, transfixed by the mad tricolor of the raptor's brilliant hues. Its crescent-shaped, bluish-gray rapier of a beak was honed for business, and its avian regalia was topped off by a flat coal-black feathered cap. EL pointed out its mate nearby on the limbs of an old, dead mesquite tree. He said that if a male and female were perched together, then you can bet that there's a reptile of some repute in a hole under that tree. Working in unison, one bird drops to ground, teasing and tormenting its victim, acting as the bait while the other watches and waits above. In time, the snake becomes exhausted from striking. Once it can no longer coil and strike, the assassin drops from above and drives its talons home.

Predators and prey: I learned their ways from EL, who understood animals better than he understood most people.

EL also taught me about a somewhat more primal pursuit: the thrill of the chase. Houndsmen are a breed apart, and EL loved that particular blood sport. Part hunter and part dog trainer, a houndsman uses the instincts of his canine companions to chase bobcats and mountain lions through thickets and arroyos where no self-respecting sportsman would ever choose to venture. The adrenalin rush is high octane. From the moment EL unleashed the hounds to the final sickening death gasp of the cat, he was in heaven.

On one moonlit night with only the faint beam of the flashlight for a guide we stalked the gravel hills of the West River. Then through the deep whitebrush thickets of the South Mott we followed the lonesome howling of the hounds as they stuck nose to trail. Each turn elicited a new howl: the staccato bark of John, the lead redbone, melding with Rufus, the black-and-tan. There were redbones, blueticks, and a bulldog that was there to do the catching. Different breeds, but they spoke the same language to one another, and to us.

One night we took the hounds to a rough and rocky section of the ranch that bordered San Ambrosia Creek. It had been a wet fall, and the water rose well up into the Carrizo Cane, leaving just enough mud at the crossings to make the tracks of a huge male bobcat visible. Shining his flashlight over the placid green waters of the San Ambrosia, EL scanned the banks for the telltale yellow-gold glow of cat eyes staring back. Seeing none, he knelt at the track and angled the light so he could study the imprint closely.

"It's a tom, all right, and a big 'un. You go up the bank and get some wood and we'll make us a little fire where we can sit after we let the dogs out and wait while they run his trail."

I dutifully took the flashlight, gathered enough small pieces of mesquite for a fire, and put some dried grass underneath. Then I called EL to bring his Zippo lighter. With the blaze lit and the mesquite popping, we sat there and listened to the painful moans of anticipation coming from the hounds, still caged in the back of the truck. EL settled himself to wait. "Let's just us sit here a spell and let those dogs get to thinking about how much they hate that cat."

After a few minutes of almost intolerable canine anticipation, we unleashed the hounds. Big John hit the ground running, circling the fire before zigzagging to and fro and then zeroing in on the tracks in the mud. The pack followed and the chase was on. We waited for our cue to close in.

"Where do you think they are now?" I asked. EL held up his hand for silence. We could faintly hear the far-off baying of the pack as they ran in hot pursuit of a scent they could almost feel. Just as I was ready to ask one more time where the hounds were, EL said, "Shhh." Big John's plaintive howl had shortened to a staccato, deep-barking yelp that signaled he'd picked up a hot scent. The Catahoula and the Walker chimed in.

EL stood, kicked a pile of dirt onto the little fire. We got in the truck. Now it was our turn to follow, not a scent, but a sound. Every few minutes he would kill the engine and lean his head out of the window to listen. Winding along old, abandoned roads and down *senderos*, we crept closer and closer to the dogs. Suddenly the baying and yelping stopped. Instead, we could hear the terrifying growl and the snapping and popping sound of bare teeth as the dogs lunged at their prey. Then there was a low moaning hiss, followed by a menacing, guttural meow.

EL aimed his special hunter's flashlight, a huge eight-battery chrome affair, toward the sound. We could see the whitebrush bend down as the cat backed itself deeper into the thicket with every lunge of the hounds.

At six foot seven inches EL was not exactly well proportioned for crawling into a whitebrush thicket to dispose of an enraged thirty-five-pound bobcat. The cat seemed to know his number was up and had no hesitation about taking on any and all comers.

"Sunny, I'm too dang big to go a-crawlin' around in there. I'll just get tangled up. I think you better take care of that cat."

Reaching behind the seat EL extracted a three-foot Mexican machete with a hard rubber black handle and a blade that had seen better days.

"Here, let me put a little edge on this and you go in and give him a few whacks."

I couldn't help but ask the obvious question. "EL, don't you have a gun?"

A grin spread across his leathery lips as he leaned out the window to spit a thin brown stream of Beech-Nut to the ground. He took a small file off the dashboard of the pickup and then, slowly and with great deliberation, began to work the edge of the blade.

When he handed it to me, I took it. I entered that thicket crawling on my belly and cradled that sharpened machete for dear life. I inched my way toward the snapping jaws of the hounds and snarl of the cat. EL went back to the truck and switched on the lights hoping I could get a fix on the quarry and start swinging the blade.

When the lights came on, the bloody carnage before me almost made me turn and try to run. The thicket held me fast. The cat was a big male, and he had swiped a claw over John's muzzle, shredding it and exposing the bloody lip and gums of the half-mad hound. With every swipe, the lip flapped wildly, slinging tiny droplets of blood over the brush.

Kicking the brush down as best I could, I rose to my knees and started swinging that machete. The dog's life hung in the balance.

But John had it easy compared to the cat. It was penned on all sides by the dogs and was madly lunging away as his tormentors edged closer. It hardly seemed fair, but it had to be done. I raised the machete and struck him repeatedly about the head. He growled and let me have a dose of claw and jaw while the machete did its grisly work. Fear and adrenalin fueled my blows.

I kept hacking at the skull. The blade cut well at first, then dulled. That didn't stop me. At a certain point, something had taken over. Call it blood lust, call it what you want. I was in the present moment with nowhere else to go.

Finally, I managed to strike a lucky blow just behind the ear and the beast stumbled back, giving John just the angle of attack he needed for the death grip around the neck that ended it all. Convulsions followed by a dying quiver.

Silence. No movement, by either of us. I rose and went to look at the cat. What I remember most were the deep gashes on that perfectly beautiful face—how crescent curves oozed blood and fur. Then I dragged my bloody prize back the way I had come.

EL was beside himself with pride at what his little man-boy had done. I felt like some adolescent Maasai who has just slain his first lion and stepped across the threshold into manhood. In truth there was little honor in such a barbarous act as mine. But the meaning of that death is imprinted in me forever. In the last seconds before I struck the decisive blow, there was just me and that bobcat. Both of us faced two fears: fear of each other, and the greater fear that we would not measure up. Or maybe that's my personal fear. One-armed men and boys who have been shamed need to make a mark in order to be visible in the world, especially in sight of other men. That's the hindsight view. But for a long moment that night, neither the cat nor I knew what the outcome would be. We were both desperate and alone, fighting with a primal adversary alien to us both.

SIX

Lessons

Growing up on the San Pedro Ranch gave me an unbounded education that could never be matched in the confines of a schoolroom. Hunting and fishing were requisite activities for any boy growing up in Texas, but there was a different kind of hunting I treasured more than anything. I searched for arrowheads, for symbols etched in sandstone—anything the Coahuiltecans had left behind. Walking down by the creek, pellet gun in hand, I would pause and look at these remnants of a people from the past. Running my fingers over the symbols and seeing the tiny lichens that had colored the lines over the past three hundred years, I would imagine their world and wish that I could live as they had.

One day in early spring, when I was about ten, a cowboy named Santos and I were out at daylight for a ride to look for strays that had been missed during the roundup the day before. We rode silently side by side. He reined his horse to a halt and dismounted, then knelt down and picked a brilliant scarlet wine cup from the sand. He led it aloft for me. The sweet smiling face of Santos, his creased and sweat-stained cowboy hat pushed back to the crown of his head and a wrinkled smile on his lips, holding the fluorescent little flower to the sky—that image has remained with me for half a century.

From the backyard of my parents' ranch house I could look out over the white stucco wall that enclosed our home and gaze at what our family called "the big tree," a live oak 150 yards distant. Dark green leaves that made a canopy thirty yards wide and a four-foot girth set this landmark oak apart. No other tree on the ranch even came close in terms of size. Its majestic proportions dwarfed the understory blackbrush and cactus that surrounded its perimeter. The shadowy confines protected all who sought refuge from the blazing sun. It was well over three hundred years old, with gnarled, dark brown bark that scraped your belly raw when you shimmied up its main limbs. A boy could disappear in the protective shelter of its ancient arms. Deer, javelina, and bobcats would all assemble and walk beneath me if I stayed still. It was as if I had disappeared and had entered a secret dimension that was anchored by the ages. No wonder the Druids paid homage to the oak.

That tree was the reason my parents decided on that particular spot for the building of their new home. I would daydream and nap in the woven hammock that my father hung between two of the lower branches, a comfortable addition to the big tree that lasted until a two-thousand-pound Polled Angus bull decided that he too should avail himself of the hammock and reduced it to a twisted mass of twine.

The headwaters of San Pedro Creek were a scant hundred yards from the tree. Between the house and the tree lay an artist's palette of Carrizo sandstone. Glyphs were etched into a soft and friable rock, open to a boy's imagination. The lines quantified mysteries beyond my reach. Some I recognized: the unmistakable etching of the underside of a turtle's shell, perfectly segmented and deeply incised. Turtle brings the rain. I decided that a certain arrow carefully carved into the sandstone and pointed in the direction of the creek could only mean one thing: "This way to water."

◆ ◆ ◆

I may have been free when I was riding my horse, paddling down the Nueces, or hunting for petroglyphs, but inside the ranch house, things were getting cramped. My parents' marriage was beginning to unravel, and although I didn't exactly know it, I could feel the walls closing in. The oak tree became a good place to escape to.

I think my parents expected me to be trouble, and I did not disappoint them. They tried to head it off at the pass. They enrolled me at the San Antonio Academy, a military school renowned more for its discipline than its academic excellence. Above all, the school had a reputation for taking on hard cases and beating a little sense into them. I gave them plenty to do while my parents fought their own battles at home. We were living in the Vietnam era: hazing and corporal punishment were still accepted means of communication, and the teachers would regularly beat us with wooden boards. The one sliver of humanity came from a geography teacher who transported me to a realm far from mundane lessons and the constant shame and harassment of teachers and upperclassmen. He read from Richard Halliburton's exotic travel writing, and as we listened we conjured up images of Halliburton diving into the sacred cenote at Chichén Itzá to replicate a virgin's sacrifice to the rain god.

I was lucky not to be of age for a real war. But I was close to

someone who was. His name was Juan Cervantez Jr., but everybody called him Tiny. He was anything but. His daily uniform
consisted of a clean white T-shirt and pressed Levis. He weighed
in at well over 280 pounds, and his five-foot, seven-inch frame
just barely gave him enough height to make it past the army in-
duction limit when he was drafted in 1966.

His parents were the cooks at the bunkhouse, and I never
saw Tiny without a smile or one of his mother's flour tortillas
clutched in his hand. He was always at the ready to help anyone
who needed it. So when he was drafted at the age of twenty-two,
he didn't resist. He was shipped off to boot camp and became my
tether to that war world away, so distant from the fantasyland
comic book rack at the Broadway Pharmacy. Just before he left,
he took all the money he had and put a down payment on a shiny
new 1966 Ford Mustang convertible. He left it parked under the
equipment shed sandwiched between a gooseneck cattle trailer
and the windmill truck. Bright, candy-apple red with a white top,
it was a sexy symbol of youth and vigor.

Boot camp shaved off more than eighty pounds from Tiny, and
by the time he landed in Saigon he was physically transformed.
His spirit remained the same. In one of his letters home to EL he
said that combat was not too bad: "It's just like playing cowboys
and Indians except that the guns and the bullets are real."

On January 10, 1967, three months after his arrival in country
and seven days after his twenty-third birthday, Tiny died in combat. The letter from his commanding officer said he was shot in
the leg by a machine gun, and as he fell, more bullets cut into his
chest, killing him. That red Ford Mustang stayed parked under
the shed for another two years. His parents kept it as some kind

of monument to a son who had been nothing but happy while in the company of others and lost his life ten thousand miles from home. Over time a thick coat of dust stirred up by the incessant hot winds of summer covered it. Trucks came and went through the caliche-covered courtyard, passing the Mustang that sat there sad and lonely, waiting for the man who had left the San Pedro as a boy. I remember my mother looking at that car one time and saying, "Why on earth did he buy that expensive sports car when he knew he was going to Vietnam?"

The VFW post in Carrizo is named in his honor. He was one of only a handful of combat casualties from Dimmit County during the Vietnam War. Several years ago I went to the Vietnam Memorial in Washington and found his name on the wall, but it was the car he wanted to return to and drive that served as the greatest reminder of his passing.

◆ ◆ ◆

In 1977 I was adrift and in and out of college. I landed at the ranch, hoping to find a roof and a meal. I found both; and at what we called the Lion Tank, I learned another lesson in life and death.

My father and stepmother had flown off somewhere for a few days, and I was alone at the ranch. I heard a frantic knock at the door, and when I opened the screen there stood the couple whom my father employed as cook and housekeeper. Their English was not so hot to begin with, but their tears told me enough to let me know there was big trouble. We jumped in a pickup and raced to the far western end of the ranch. As I crested the hill above the Lion Tank, my heart sank. There in the middle of that half-acre pool was my old faded blue fiberglass canoe, its bow protruding above the waterline.

The cook and housekeeper had taken the day off to go fishing and had decided at the last minute to bring along one of the ranch hands, who was known only as Crazy Antonio. He had a seemingly insane sense of happiness that seemed to suffuse every cell of his being—hence the moniker. The three had been fishing in the tank, but at some point, they said, Antonio had fallen overboard. They had run to the ranch for help and brought me there. Together we threw a line to the canoe and pulled it to shore before fishing for the body.

The sheriff, Doc Murray, who was far from being a doctor, arrived. From a leather bag he produced two eighty-pound-test monofilament lines with four-inch grappling hooks attached. We swung those lines in rhythmic arcs over our heads and let them fly. First, from one side of the boat, and then from the other, we watched the heavy hooks splash then disappear beneath the surface. The lines grew taut as the weight of the grappling hooks bumped along the muddy bottom of the tank. Just as I was pulling my line in for another throw, I felt the sickening resistance of pressure on that line. I pulled hand over hand and called out that I thought I had something. The next thing I saw was Antonio's green-and-white-checkered western shirt as the hump of his back broke the surface.

By now a few more hands had arrived from the headquarters, and they helped us lift Antonio into the back of a pickup. His body was stiff, his face an ashen gray. His long black hair was matted down over his forehead like a kind of helmet. In the background behind the pickup stood the couple who had brought him there, shifting from one foot to another and staring at the ground.

The next day our head cowboy, Romeo Boone, decided to do a little checking around behind the dam at the tank. There, carefully tucked under a thick stand of guayacan, were two of my father's fancy French wine bottles, both drained. When Doc Murray questioned the couple, they eventually admitted not only to being drunk but also to accidently causing Antonio's death. It seems that they knew Antonio loved to fish but was afraid of the water because he could not swim. When they were in the middle of the tank, their drunkenness got the better of them, and they decided to torment Antonio by rocking the boat side to side. Soon enough they rocked too far and Antonio fell in, capsizing the boat while they swam to shore.

Finding him was hard enough, but it was the immediate aftermath of this tragedy that I remember most vividly. As we were loading the body into the pickup, I saw a cloud of dust on the road. I could hear the sound of rocks hitting the frame of the vehicle as it tore toward us. Antonio's twelve-year-old son leaped out of an old red Ford van that served as the ranch school bus. He could barely see over the dashboard, and he had crashed through four gates to get there, but he was not to be stopped, or comforted in any way. He wailed inconsolably, clutching and tearing at the soggy shirt and jeans of his dead father, who lay in the bed of the truck.

When my father returned, he thanked me for being there during the tragedy and sent the local Catholic priest to try and console Antonio's son. Eventually the boy returned to Mexico to live with his widowed mother.

I couldn't forget the image he'd left me with: a son trying to hold on to a father who has left the world without saying goodbye.

◆ ◆ ◆

Just west of Del Rio, Texas, is one of the most vivid and enigmatic series of rock petroglyphs and pictographs in North America. In an area roughly three hundred miles in diameter where the Rio Grande, Pecos River, and Devil's River converge lie some of the best-preserved and oldest indigenous artworks in North America. The most ancient of these pictographs date from four to five thousand years ago. Then as now, if it weren't for the rivers, people would never have been able to live in such a harsh world. Their paintings show their gratitude to the natural world that gave them guidance, comfort, and the power to sustain themselves over centuries, working in concert with a world where nature dominated their every move.

Each image gives the illusion of living in another dimension. The people who lived here created a world of their own choosing, where there was no firm boundary between the here and now and the great beyond. They are lithic declarations that amplify the intersection of the sublime with the metaphysical.

Rock art is much more than just an anthropological romp. It is a passageway to the past. Fantastic visions beheld by small extended family groups and nomadic journeyers painted with a purpose. Or the whimsical stylized bison painted in red monochrome ochre, dancing on its hind legs with horns and tongue extended. At a shelter on the Devil's River is an anthropomorphized shaman panther hybrid. On his extended hand sits a bird, symbolizing the flight of the soul and life in the upper world. It is a universal symbol found the world over, from the cave at Lascaux in France to the Devil's River, and beyond the flight to the heavens it keeps us grounded in our everyday world. Every little bump, every flake of color leads you on a story path that is yours, until interpretations become secondary, and it is the moment you are here that takes on meaning. The images serve as touchstones anchoring us to a past where we long to go but most likely could not endure.

◆ ◆ ◆

After earning an undergraduate degree in 1985, I began what I considered to be my informal advanced degree. It had taken me eleven years to finish college. But once I got a taste of learning in a classroom, I was all in. I slugged through a master's degree in history and emerged proud and unemployed.

It became clear to me that if I became a teacher, what I wanted to pass on to students had more to do with the past than about what to expect in the future. I began as a substitute for the San Antonio Independent School District and applied for a full-time teaching position on San Antonio's west side. I was determined to bring the lessons of indigenous peoples into the classroom of the twenty-first century. The children of the inner city, I was sure, could greatly benefit from the example of their ancestors.

I begged the superintendent for a job. I sat across from him in his office and told him that these kids needed to connect to who they *really* were. They weren't Mexican Americans, I said. They were Native Americans, descendants of the proud tribes who once held sway over South Texas and had the wherewithal to survive and thrive in an inhospitable land. Since European invaders had taken that land from them, I thought it was high time that something be given back.

The superintendent tried to tell me nicely that I had no clue what I was doing, all of which had the effect of making me dig my heels in even deeper. It didn't take long for me to realize that I was digging my own grave, and that I was the one who would get the real education. The school was on the rough side of town, the barrio and then some, replete with all the trappings of poverty, gangs, drive-by shootings, single parent struggles, and drugs. The school was often a student's last stop before incarceration at juvenile detention. From there, it wasn't far to the penitentiary.

Some of my students were monsters, but they all had names. My teaching assistant, I'll call her Maria, a proud Puerto Rican with a mean right hook, had a method for literally taking matters in hand. She had been where those kids were, only more so. She had pulled the trigger, snuffed out a life with fourteen 9 mm hollow-points into the broad muscular chest of the Crip who took her brother's life. The kids respected her. She walked the walk. So far, I was mostly just talking.

I would arrive in the morning with my lesson plans to find the class was gathered on the open-air basketball court. Jesse shot baskets, and his brother Mando looked forlorn and crazy with anger in the corner of the court. His hair was slicked back, thick and greasy, and he had the telltale smooth craters on the left side of his cheek that marked his rite of passage into the tribe, brands left by a cigarette stubbed out on flesh. Whatever I had suffered in school seemed tame by comparison.

I took the class on a field trip to Mitchell Lake, the historic trash dump for the city of San Antonio since its founding. The lake had been newly transformed by the Audubon Society. The sludge had been treated so that migrating waterfowl could feast in the shallow pools again. As we watched, puddle ducks popped short gulps of water through brown bills and a raft of white pelicans rose from the lake and glided effortlessly to the far side. Transformation of even the most unlikely people and places was still possible. That's what I wanted my students to see.

One of my students, a kid named Giggles, pulled on my shirt and gave me his toothy grin. He placed a perfectly preserved skull of an arctic tern in my hand. A black cap of tiny feathers still adorned this migratory sojourner, which had left a colder clime but was vanquished on the shores of what it must have thought was a tropical paradise. The bird had no business being down here, really, but it was following the commands of biology and instinct.

I thought that if I took my classroom to the woods and taught my students about shelter, fire, living off the land, and archaeology, they would be transformed. I could feel their hunger for learning about the people who came before them. The boys came alive when I set up soda cans on the basketball court and demonstrated the use of the rabbit stick, a piece of wood roughly the length of your elbow to the tip of your fingers. You simply look at your target, cock back, and throw, stunning the animal so you can then dispatch it by hand. But the only stunning came when I saw the terrified look on the face of the principal when he looked out of his window and saw that I had armed rival gang members with weapons and was encouraging them to sharpen their aim. The administration did not share my enthusiasm for teaching the old world, and ultimately I was brought before the head of the department and told to teach to the test rather than instructing students on how to throw an atlatl or use a bow drill to make a fire.

Trying to keep a twelve-year-old aspiring gangster in a classroom was like trying to hold back the dawn. No matter what I used—tricks, treats, or subtle suggestions about how education is the way out—nothing worked. History of their proud ancestry was gone. The present was constantly changing and uncertain. They already knew the future, and it had nothing to do with them. They didn't expect to live that long.

◆ ◆ ◆

After a time I left the school and returned to something that had always been a source of strength for me. This time I became so absorbed in the Pecos River style of rock art that I served as the director of the Rock Art Foundation, an organization dedicated to the preservation and protection of these ancient treasures. During that period I took numerous trips to show people polychrome panthers painted on cave walls, shamans in ecstatic trance with their hair standing straight above their head, and whitetail deer leaving a trail of deep crimson drops of blood across the rock, an atlatl dart wedged into the chest cavity.

The paintings had a unique effect on people. They caused us all to stop and be silent. As we stood in those shelter caves, shaded from a sun that felt like it could melt you on the spot and protected from wind that scraped the landscape mercilessly, we stood in awe before the visual record left by the ones who preceded us. "We were here," the pictures said. "This is how we lived." My classroom was now a shelter cave where rites of passage were conducted and ceremonial sites lent a protective aura to the harsh world of the Trans Pecos.

To clue yourself in, to go deeper, to step off the edge and see whether you take flight or freefall to the rocks below, to make a concerted effort to change without really knowing you are about to do something that will set you off in another direction: that is what happened to me when I enrolled as a student in Tom Brown's Tracking School. A close friend had suggested this class to me, and I placed myself at the feet of the man whom most consider to be the country's master tracker and survivalist. It had a boot-camp type of atmosphere that grated on me at the beginning. But you learn a lot about yourself when you eschew the comforts of home and hearth for a sleeping bag and day-old granola served communally in a drafty old barn.

Deep in the pine barrens of New Jersey on the Delaware River I joined Tom and seventy-odd fellow survivalists. We had come to learn the fine art of becoming who we used to be. We slept in an old barn on bedrolls, severed from our high-tech lifestyles. Our cell phones and credit cards were replaced by rabbit sticks and bow drills. We found water with a witching rod and built a debris hut out of twigs and detritus we gathered from the forest floor. The experience was as much an education in philosophy as in acquiring new skills, and I could feel myself absorbing the lessons as my skills improved.

The great test came in making fire. Prometheus must have chuckled at my struggles, watching over my shoulder as I hacked away at a block of wood until a crude bow drill emerged. A twisted cord of jute for a string, a block to hold the drill, and you start the process: the slow and steady friction back and forth, back and forth, until that first faint whiff of carbon wafts upward and an ember starts to glows. You add cordage in the form of fluffed-up jute, building the flower up slowly and with great care. Breath and fuel, the slow arc of time, the reawakened art of patience and resourcefulness: this is a moment locked in memory.

The beauty in learning to track is that it forces you to slow down and pay attention. You become lost to the world as you create the story of what you see in front of you. Everything and everybody leaves a track of some kind. It may be physical, as when an animal disturbs the soil, the imprint revealing the size, sex, or attitude it had when the mark was made. Everything leaves a sign. Our task was to learn how to find and interpret what we found: the way a deer plants her hoof on the trail, the length of stride she takes, and the imprint of her hind foot on the outside where she has laid a front hoof all reveal her gender and purpose. The age of a track can be determined by how much surrounding soil has fallen into the track itself. It is a secret knowledge, as simple as it is logical, requiring only the willpower to stop, look, and slow down. A rear hoofprint that falls outside the front hoofprint is made by a female, because giving birth requires wide hips; the opposite is true for males. A browsed bush where a deer's incisors have cropped the tender new growth of twigs becomes a beacon of light leading to a visual retelling of what has passed here.

By the end of our four-day intensive I could make a fire with a bow drill, build a shelter, make cordage, find water, and track quarry. I was honing the same survival skills used by the people I had long admired.

The first book I remember having a hold on me was *Ishi*, by Theodora Kroeber, the daughter of the anthropologist A. E. Kroeber. I found it in my parents' bedroom library, and from the moment I opened the cover I could feel the pull of Ishi's story. In 1911 a Native American man had been living alone and abandoned in the woods of Northern California. He spoke only the language he had learned as a child. For many years, he literally had no one to talk to. A photograph of him at the time of his emergence into the white world shows a wasted skeleton of a man whose dark sunken eyes seemed to have no light, only black holes that mirrored the suffering he had witnessed. Ironically, he had reentered the twentieth century by seeking sanctuary at a slaughterhouse adjacent to his ancestral hunting grounds. A professor in the anthropology department at Berkeley "discovered" him, and for five years he was both instructor and inmate at that school. Photos of him shooting a bow and arrow and making fire with a bow drill while patiently demonstrating the art of survival have been indelibly embossed in my psyche. Here was a man who had stood by and watched as his people and his family succumbed to the violence and diseases of the twentieth century. He faced extinction, yet had the graciousness to pass on his wealth of knowledge about the natural world, a world that was quickly vanishing. To communicate he used sign language and gesture and a smile, saying more than words could ever hope to. He passed it on, telling the story of a world from which he had emerged, before he died a slow death from tuberculosis, in the confines of a university hospital where the modern world labored to save a remnant of the past.

◆ ◆ ◆

It was time to get back to the present. I wanted to have a connection with a living tribe closest to the ranch, and I became friends with some Kickapoo Indians who lived in Eagle Pass and Nacimiento, Mexico. I went to the elders one day and told them I could offer deer hunting on my land. They talked among themselves for a time and then replied that they were very grateful. They explained to me that the deer was their intermediary to the Creator. When a deer is killed, they attach their prayers to the spirit of the deer as it rises up into the heavens. Without the soul of the deer there are no prayers, and without the prayers there is no intercession by the Creator to help the Kickapoo.

My part in this endeavor was more mundane at first: I bought them a freezer and took them deer hides. The chairman of the tribe welcomed me, I became acquainted with members of the council, and soon I was hunting with the Kickapoo.

A young man ten years my junior was appointed to be my guide and interpreter. He also became my friend. Juan Gonzalez took me to the sacred headwaters of the Sabinas River. There we swam under the canopy of ancient cypress trees and watched a world that seemed to have been forgotten. I placed an offering at the spring and then went inside a wickiup for a traditional meal of deer ribs and stew. The wickiup's walls were large mats made of tule, the cattails sacred to the Kickapoo. In the spring certain men are selected to dive into ponds where the cattails grow and cut them below the waterline. The women then bundle the stalks and take them back home for drying and weaving. In a dry year I once had them come up to the botanical gardens in San Antonio to gather tule there.

One day the administrator of the tribe called me aside to tell me how the council viewed me. "They are extremely grateful for your help. It is important for you to know that while they may not be able to show you their gratitude, they do pray for you." That statement meant more to me than anything. I felt that I had allies.

Juan Gonzalez was driving down to Nacimiento early one morning when a smuggler working for a drug cartel ran his truck off the road, killing him. Afterward, I traveled to the sacred springs where we once swam together. I couldn't help but wonder whether a hundred years from now, a child might come upon that leafy place and find joy in it—and, more important, whether there would still be water.

More than ever, the fate of the Rio Bravo pulled on me. Did I choose the river or did the river choose me? Either way, the river is a winding, disappearing conduit that gives meaning to my being and sustenance to my soul. Humans have always besieged this river. We cannot seem to stop ourselves when it comes to taking what is wild and free and harnessing its essence to serve our own desires. For we want to live and prosper, want to have the river give us what it has. The river does until it gives out, exhausted by our ambition.

What is *my* nature? I was born in a different time. The Kickapoo taught me by example how to see the land and its resources as sacred, but my roots are in a family that has taken from the land in order to prosper. Finding my own place to stand between these worlds has been a challenge, but the trail I leave is my greatest concern.

SEVEN

Cathedral Rock

It was still dark in the last hour before dawn when I woke. Ink dark. A crescent moon accompanied by stars set like tiny precious jewels into the black void of space slipped beneath the western horizon. Infinity defined. It felt like the kind of dark that can give cover, and it wrapped itself around me like an old, frayed blanket. It held me close as I balanced on a tightrope stretched between terror and tranquility.

I couldn't sleep. Nightmares, maybe, like the frightening vision of an eighteen-wheeler tanker truck with its grill opened wide, a drunken steel predator careening down a ranch road toward me. In this recurring nightmare, I watch helplessly as the truck crashes into the old red wooden swing gate at the Number 5 windmill. The peeling, faded red planks of lumber splinter into shards that launch skyward, and the rusty chain and O-ring that has been there since before the war snaps, its links flying in all directions.

I was on a mission. I knew where I had to go, and it wasn't too far away. I tried to pay attention to everything as I walked slowly across the ranch. My senses would soon be filled with what was actual as the gradual liberation of light gave new life to all that surrounded me. But at that moment my eyes were glued to the ground.

My destination was Cathedral Rock. On the flat plains of South Texas, it's a notable landmark. I made that journey not for the pleasure of the view, but because there was something I *had* to see.

A dreaded phone call had set me in motion. Without my knowledge a decision had been made, one that had been suspended over my head like the thin, finely honed blade of the executioner's ax, and I didn't see it coming until my neck was on the chopping block. From Cathedral Rock I thought I might be able to glimpse my fate, or at least get some perspective on the forces that threatened all I hold dear.

As a boy, I rode a little bay Mexican gelding named Cacahuate, which means "Peanut," to that remote corner of the ranch. His restless nature would not tolerate an extended viewing of the rock, but the sight of it was enough to set the hook of its mystery deep inside me. Back then the place was off-limits, corralled by a neighbor's fence. In midlife, with permission, I approached this monolith on foot in the manner of a supplicant, as a seeker, a son, a brother, but also a husband, a father, a rancher attempting to be a good steward of this land that had been handed down to me. Above all, I was a man arriving with questions. But how do you talk to a rock?

The oil company man wanted me to meet his survey crew the following week. They were coming to my pasture to drill ten wells on two five-acre pad sites, and I couldn't do anything to stop it. But I wanted to get a clear view of what had already happened to my brother and sister's land, south of mine. On the other side of their fence was the rock I was determined to climb.

The light grew by degrees. I paid less attention to my feet and more to what was ahead of me, but I moved slowly, for it was early spring, and that meant rattlesnakes were on the move. They had been snug and warm all winter in their dens and burrows, luxuriating in their torpid state of suspended motion as their cold-blooded bodies slowly digested the pale yellow fat they had stored all year. Waking to a ravenous reality, cranky and hungry like some hungover frat boy the day after Mardi Gras, they would strike at any heat source that even remotely resembled prey.

The biggest rattlesnake I ever saw was more than eight feet long and weighed ninety-seven pounds and four ounces. It had thirty-two rattles, and the diamonds on its back were the size of playing cards. Game warden Herbert Ward had killed the snake as it crossed the road near Pendencia Creek. He said he knew it was big when the rattler's head got to the far side of the road and the tail had yet to emerge from the brush.

Being struck by a snake that size would be like taking one in the solar plexus from Sonny Liston. The girth was the size of the trunk of a small mesquite. When they slit open that pre-historic reptile's belly, two slightly decomposed baby jackrabbits were halfway down its digestive track. Above them reposed the slimy gooey-feathered mass of the snake's final morsel, a male mockingbird whose ebony black eyes were now forever frozen, as if locked in disbelief that it had come to this.

A mere photograph of the snake was enough to keep most people in their cars. And while the length was impressive—with a six-foot cowboy holding it up, there were still two feet of snake coiled at his boots—it was the girth of this serpent that got your attention. At its widest point it measured nine and a half inches. Almost one hundred pounds of pure muscle, capable of driving twin hypodermics the size of small fountain pens deep into prey that never had a chance.

With that memory fixed in my mind's eye, I scanned the rock slowly, trying to quiet an imagination gone astray, then rose to my knees and took a step. The pasture-walking rule in South Texas is that you only look up at your surroundings when you stop moving. It's a habit you quickly acquire after your first encounter with Charlie-No-Shoulders.

I walked toward the final boundary of the fence and found the worn passage of a deer trail under the barbed wire. The telltale patches of dark gray hair follicles suspended from the rusty barbs floated gently back and forth in time with the morning breeze. Falling to my belly, I inched forward at eye level; I stood up and began to walk again. With each stride the rock grew in size until I stood at the base. Slowly, reverently, I felt the coarse sandstone with the tips of my fingers. Here, where words fail, touch can tell.

Its color is ghostly, flecked with variegated shades of lichen that adorn its sides, forest green giving way to pale gray and then to black. The lichen are here because they have found a home that will have them, advancing imperceptibly to grow when there is moisture and receding to tiny brittle flakes that do their best to cling to their foundation as heat and dry winds scour the stone's surface. For millennia this land has endured despite its limited capacity to reproduce, self-regulated by whatever might come from the sky above, a hydrologic cycle that came and went and circled back with an irregular wobble that humbled all who entered its realm. Life adapted. Little rain meant a profusion of mesquite beans, scorching temperatures, and animals like the jackrabbit, which evolved with long ears to dissipate the heat.

I chose a circular path to take in the magnitude of what I was seeing around the base, studying indentations and fallen pieces of smaller rock until I found what I was looking for. There are seven wide and deep grinding holes incised on a narrow, waist-high ledge. They have a language all their own. For as the rock was slowly and painstakingly abraded and the holes deepened, they took on a different force and tone: the deeper you go, the longer you last. And in the world of the native peoples who first set their eyes on this rock, longevity was somewhat ephemeral. Cathedral Rock is the height of a very large four-story building rising over an open sea of grass, cactus, and brush. In geologic parlance it is known as a pillar, a wholly inadequate term for such an imposing anomaly. It was formed more than 35 million years ago, a remnant of a shoreline that bordered the sea during the Eocene epoch. What kept it from eroding was that it was different from the surrounding stone. As the saltwater flowed through it, a stronger rock was exposed, and sand changed to quartz. An alchemical transformation ensured that this pillar of stone would stand.

You have to adapt to such places, not the other way around. But the dust and the din, the clamor of pipe and the raucous cacophony of chain beating against iron propelled me toward that rock refuge. Here I could be surrounded by the reality of time and money, but I was also held safe in the shelter of another time and place.

Halfway up the rock, a wide flat ledge unfolds in layered patterns of undulating sandstone. On the largest of these slabs are the names of an entire squad of National Guard cavalry troopers—Troop L of the Illinois State National Guard—and a date: November 2, 1916.

I wonder what those Corn Belt farm boys and Chicago slum kids must have thought as they stood on this rock, sent into the middle of nowhere to protect their country from the revolution across the river. Mexican bandits were the main threat, crossing the river to loot and pillage as far as they dared. These young guardsmen in scratchy wool uniforms, lean and in their prime, were likely bored stiff by a life of dull routines and drilling, but a midnight raid by the light of a full moon pierced their world with the power of the vengeful and the righteously indignant.

On June 19, 1916, President Woodrow Wilson mobilized the entire National Guard and sent them to the Mexican border. Pancho Villa and his gang had been raiding Texas ranches for livestock and then trading the cattle back for rifles and ammunition. For several weeks after the arrival of the guard, the river country was abuzz with rumors of a large-scale raid by a band of disgruntled Villistas. They had nothing to lose in this no-holds-barred border warfare. When a *New York Times* reporter tried to explain the Geneva Convention to Villa, all he got was a furrowed brow and a two-word response: *¿Por qué?* Why? Villa knew full well that personal enrichment was a short ride away, and safety a sure thing, with a fast horse and a speedy retreat across the shallow river. After each raid and subsequent retreat across the river, there would be the inevitable retaliation by Texas Rangers or angry ranchers.

Three years before that inscription, not five miles distant from Cathedral Rock, Dimmit County had its most notorious encounter with the Mexican Revolution. They were doomed from the start, those revolutionaries, a ragtag confederation of Marxists, Zapatistas, and Wobblies known as the "Red Flaggers," born of desperation and cemented by anger. President Wilson had recently invoked the neutrality laws that were designed to place impediments on just such groups as this, but in the wilds of the Rio Bravo, a law is a long way from its enforcement.

The revolutionaries' plan was to open a northern front for Zapata, to be launched on September 16, the day Mexico declared independence from Spain. Their flag bore the prophetic words *Tierra y Libertad*, a slogan borrowed from the newspaper published by the anarchist Flores Magón brothers. The revolutionaries smuggled German Mausers, cases of dynamite, and various other munitions by train to Carrizo Springs. As the wagon bearing the arms left town, it was spotted by Deputy City Marshal Candelario Ortiz, who shadowed the group until they discovered him and took him prisoner. They loaded him down with bandoliers of ammunition, and when he collapsed from exhaustion they shot him dead. Likely they viewed Candelario as a turncoat because he was a Mexican enforcing the white man's law in a land that they saw as theirs.

Word was telegraphed to Fort Duncan upriver, and a contingent of cavalry on motorcycle and horseback started downriver. Meanwhile, a posse from Carrizo Springs rode out to trail the gang. It was September and hot, a four-inch rain had deluged the area the night before, and the earth turned to muck while the humidity suffocated the posse and the revolutionaries alike.

When the revolutionaries reached the San Ambrosia Creek, which flows through the West River pasture of the San Pedro, the water had risen so much that they thought they had reached the Rio Grande. Once they had crossed the creek, they killed a deer and built a fire, believing they were safe in Mexico. The posse and the cavalry saw the fire and charged down on the unsuspecting men with guns blazing, killing one, wounding two, and capturing the rest. Sheriff Gardner and Deputy Gene Buck were hailed as heroes, as was Lieutenant Terry de la Mesa Allen, who would become George S. Patton's second in command during the Second World War.

Buck sent a film crew to Carrizo Springs in 1914 and produced a silent film about the incident, starring himself and other members of the posse. As for the revolutionaries, they were sentenced to prison terms in Huntsville, only to be pardoned in 1928 by Governor Miriam A. Ferguson, who declared their imprisonment unjust because their actions were an act of war.

A few years ago I set out with my father and a friend to locate the long-forgotten battlefield of these insurgents. I knew the approximate whereabouts of the firefight and was determined to see what I could find. After hours of searching, I retreated to a small ravine and sat down in the shade. When I got to my feet again after cooling down, an unnatural object caught my eye. Bending down, I unearthed an unfired rifle cartridge. The short stubby round was a Winchester .351, an exact match for the rifle used by Sheriff Gardner. I had found my object of desire, a link to the past that has connected me and kept me questing and wondering to this very day.

◆ ◆ ◆

Tierra y libertad: the battle cry of the disenfranchised revolutionaries. They had each other. I related to a group who knew they had a right to what was lost, knew the injustice brought on by years of subjugation, dominated by a force as pernicious as it was mysterious.

There is a wonderful quote by Gen. Porfirio Díaz, the Mexican dicator who ruled the country for more than three decades: "Poor Mexico, so far from God and so close to the United States." In February 1915 the Mexican Revolution finally gained a foothold in Texas that shook the border to its core. As a means of instilling fear in Texans and fomenting retribution against their enemies, a group of disgruntled revolutionaries who were imprisoned in Monterrey hatched the infamous Plan de San Diego.

The idea was to turn up the heat on the smoldering relationship between Anglos and Tejanos. The manifesto, named for the small town in Duval County where the plan was eventually discovered, stated that any white male over the age of sixteen would be executed. A no-quarter race war that exempted only women and children would also repatriate Native Americans to their lands and spare African Americans. The ultimate objective was to free the territory that the United States had seized from Mexico after the Mexican War.

It is likely that Venustiano Carranza, the leader of one of the revolution's most prominent factions, was behind the plot. By sending small guerrilla bands into Texas and killing Anglos and Tejanos who were allied with Anglos, he managed to create an atmosphere of paranoia along the river that permeated the border.

As a direct result of the killings, the Texas Rangers and the US Cavalry took revenge on innocent Mexicans in Texas and across the river. "Shoot first, ask questions later" became the theme.

President Woodrow Wilson eventually recognized Carranza as the legitimate president of Mexico, a move intended to bring stability to the region. The raids and depredations ceased, for recognition meant arms and ammunition could be legally exported from the United States to Mexico, which in turn meant supremacy for Carranza.

During the heyday of La Raza Unida the heat was turned up once again. The reality that the Anglo population in South Texas was outnumbered wormed its way into my father's thoughts one day in the early 1970s. Both of his Anglo managers had gone to town, and he was sitting alone in the office at headquarters. Looking out the window, he saw the cowboys sitting on the bunkhouse porch smoking and drinking coffee. Wow, he thought to himself. Now I know how Davy Crockett felt at the Alamo.

Ever since Stephen F. Austin and his "Old 300" set foot in the Lone Star State, the coexistence of Tejanos and Anglos has been unsettled. We need each other, yet history has imprinted both sides with equal parts fear and suspicion.

◆ ◆ ◆

The rock was warming with the day, and the slow southeast breeze off the prairie was steady now. Below me I watched as intermittent gusts of wind pulsed in waves, sending fluorescent green new-growth mesquite into temporary spastic fits and twists. Above me was a perfect foothold, and with an extended reach and stretch I wedged my boot firmly in the crevice, swung my leg up, and came face to face with a name I know.

"Billy, and Tommie, Ward. Family picnic, 1968" was etched in stone. When I first met Billy, the patriarch of the clan, he was standing by a swimming pool and holding a can of Lone Star beer. He was wearing a sleeveless cowboy shirt, replete with pearl snap buttons, over a pair of bathing trunks that he had probably outgrown in junior high. On his head was an enormous Mexican sombrero; on his feet, hand-tooled lizard-skin cowboy boots. A silver belt buckle the size of a small dinner plate was cinched around his trunks, holding up the rather sizable reservoir of a lifetime of Lone Star and completing a one-of-a-kind sartorial statement. Billy obviously considered this a formal pool party.

When I joined the group that had gathered around him, Billy was at the tail end of one of his favorite performances. He was regaling the boys with the legend of Billy the Linebacker, relating his time as a member of the Carrizo Springs Wildcats football team. "Course I started to slow down after a while," he lamented while patting his gut. "Toward the end of my senior year, Coach started to call me Dunlap." Taking the bait, I asked him, "Why did he call you Dunlap?" "Well, one day, he walked into the locker room after practice and said, 'Damn, Ward, your belly's done lapped over your shorts.'" Howling at his joke, Billy tilted backward on the heels of his boots, catching himself at the last moment and stumbling forward, jostling foam from his beer as he righted himself.

After the pool party we all descended on a BYOB joint out toward the river called the Pan-American Club. It was a windowless cinder-block dive, painted black with pink and violet clef notes dancing down the facade. Here Billy's brother and his wife joined us. After a brief stint on the dance floor, where Billy had tried valiantly to dance with two different women at the same time, he retired to the parking lot and the bed of his pickup. He lay there on his back, counting out loud the stars in the sky but never being able to make it past seven before they all blurred together.

Back inside the club, one of our fellow partygoers nudged my leg under the table. Her hard brown eyes were fixed on the dance floor as two scarlet red lacquered lips planted themselves with deft precision around the filter of a Virginia Slims Menthol 100. She took a long pull, leaned back, and cut her gaze to me. In a throaty whisper she purred, "Listen, Sunny, don't be such a scaredy-cat. It's my husband that is married, not me." A cold wave of weakness washed over me, and just for an instant I looked into the eyes of this dancehall veteran, who was just slightly younger than my mother. Then I bolted from the table like a frightened rabbit, retreating to the parking lot to see if I could help Billy get past seven.

The name Neal Ward is carved next to his parents, and his birthdate, November 14, 1958. In the summer of that year, just before the drought broke, my father made a home movie to capture the suffering on celluloid. In the film, a crew of cowboys was fanned out in a wide-open pasture that was nothing but cactus and dirt. On their backs they carried butane tanks and flamethrowers. They used these to burn the spines and needles off the prickly pear so the steers could eat it. In regular progression, first came the cowboys, then the steers running over each other to get to the sugary green pulp.

On the edge of the screen I could just make out the distinctive shapes of my two older cousins, Jimmy and Leo. They had been offloaded by my father's half-sisters to spend the summer in semi-incarcerated work detention under the tutelage of the ranch foreman. The contest between them was who could set the most wood rats aflame as the furry creatures ran for their lives, the roar and hiss of the butane torch serving as the dinner bell for the four-legged. Then came the most surreal sight of all: wild whitetail deer, looking like emaciated refugees from some drive-through animal park, trudging slowly behind the steers and mopping up what precious little green remained.

A month later it rained, shortening memory in all but the oldest and most hardened of our inhabitants.

Cathedral Rock was once called Mendoza Rock, honoring a Spanish soldier who, during the early explorations of this land, spied this landmark from across the river and thought, "There, that's where we want to go": *más allá*, farther on. The rock seemed like the only place where I could get at least a glimpse of the forces that shaped a lifetime of wanting, striving, pleasing, failing, all from the grand height of eighty-five feet.

My grandfather elevated himself by riding across this same ranch on horseback, for that was all the vantage point a man needed. Being on horseback tends to allow for healing and reflection. In the words of my father, "Nothing does more for the inside of a man than the outside of a horse." Maybe that was why I had to climb higher, to even begin to see what was left of the empire my grandfather had tried to create.

On a small horizontal ledge off to one side, carved with great care in big block letters, is the name John Robinson. John and his wife Marie were my grandparents' domestic helpers for fifty-two years. They had witnessed a lifetime of family fun and trouble, never swerving off the path themselves.

They had left Shreveport, Louisiana, along with my grandparents, during the winter of 1931. My grandfather, who was in the oil business, had recently fallen out of favor with Governor Huey P. Long. It seems the Kingfish had decided to impose a special excise tax on independent oilmen, and when Grandfather did not play ball, there were no more drilling permits. After John found out that his boss's fortune had turned, he made sure that he was going to be around to carry the bags to Texas.

As I stood by the rock and looked up at John's name, I remembered his face. He was coal-black, the color of the cast-iron frying pans he used to cook with. His smile creased and wrinkled every inch of that face, but it is his hands I remember the most—the cracked and weathered hands that were born to hard work and even harder times, hands whose first work for wages was to wrap themselves around the hard hickory handle of a nine-pound hammer and drive steel spikes into creosote crossties, anchoring the polished rails of the Shreveport, Gulf, and Western to the earth. He was a bull of a man, short and compact, moving through life like the freight train he was.

It was always curious to me why John would have left the confines of the kitchen and journeyed to this remote corner of the ranch to climb a rock and carve his name. It was not his nature to venture out, but when I found the answer it made perfect sense. After John died, my grandmother told me the story of their journey to Texas and how John was initially refused a driver's license because he was illiterate and could not sign his name. In her kind and considerate way, she drove John down to my grandfather's office one day to remedy the situation. There his secretary had laid her lily-white hand over that massive black one and patiently traced his name with a large #2 carpenter's pencil for the better part of the afternoon. When he got the hang of it he signed everything he could find, including Cathedral Rock.

John and Marie Robinson had been fixtures in my father's life since his birth, and my father loves to tell the story of his birth. According to him (mainly because there is no one left alive to contradict him), he was born on the coldest day Shreveport had ever seen, January 19, 1930. The only way to keep him from freezing to death was to wrap him in a blanket and place him in a dresser drawer on top of the furnace. The visual I get from this is very strange indeed. Why would the only son of one of the richest oilmen in Shreveport have to worry about freezing to death?

They made their exodus from Huey Long's land of wealth and plenty—wealth for a few, plenty of misery for most everybody else—and drove to Texas to start over. My grandfather had been born there, in a little wide spot in the road outside Gonzales called Thompsonville. He was returning home with a pocket full of money and his beautiful wife and precious little boy in the back seat of the Packard, John Robinson at the wheel and Marie the cook, whose scowl at everything and everybody were her everlasting trademark but who earned redemption with her fried chicken and lemon meringue pie.

Gone to Texas on a roadway lit by the incandescent flaring gas wells of East Texas. "All that light," my father once told me, "it was like it was midafternoon at three in the morning." In the early 1930s a man with cash could name his price, and my grandfather ended up with two ranches. He managed to meet the note payment on Llano by turning to what was there, native pecans and wild pigs, harvesting the nuts from the river bottom and trapping the pigs to sell at the sale barn. If a man had a little get up and go, he would get where he wanted to be. It was all here.

Two ranches, a big new house in the best neighborhood in San Antonio, a beautiful adoring wife. Grandfather started playing polo and traveled back and forth from ranch to ranch, making money with the money he had made. He watched little Hugh Baby ride a horse and learn to swing a mallet. In 1939 he gave my father a Winchester rifle. I have a photo of him on my wall. He looks like a miniature Roy Rogers, his Stetson Open Road hat cocked back in pride, lever action at the ready, standing between two massive whitetail bucks, hanging and dripping blood onto the polished concrete of the driveway on Encino Avenue.

Hugh Fitzsimons Sr. bought the San Pedro ranch from the Frost Bank in 1932 at the bottom of the Depression. It consisted of approximately 33,000 acres in the form of a Spanish land grant that dated back to 1812. He added another 20,000 acres, and by the time he died it was one of the largest landholdings in Dimmit and Maverick Counties. Consisting of vast grassland savannahs mixed with thick brush and mesquite, it was a rancher's heaven. Owning his own ranch was a dream of my grandfather's ever since he threw his leg over the back of a horse and penned his first wild steer.

But the original dream can only be sustained if the following generations keep it alive. In order to dream, you must look inside yourself.

Ranchers are known to hold on to their dreams, perhaps even to their own detriment. A friend once told me the story of how a neighbor, faced with the foreclosure of his ranch, sat down in front of the banker and said, "Mr. Gross, you might make my children paupers if you foreclose on my ranch, but I'll make yours orphans." They worked it out.

◆　◆　◆

I started back on my climb. I already knew that from the rock's summit I would scan the scarred remains of what used to be a pasture. But I wanted one last bit of reverie to remind me of another time, so I turned and faced the north, following the fence line of the East Starr pasture, where the width and breadth of the grasslands undulates and is easy on the eyes.

• • •

I didn't need to be reminded of the scraped caliche roads, the gleaming oil storage tanks, and the incessant vibrato of diesel engines, all of which had eclipsed the last remnants of what I have always seen as sacred. It was inevitable that even more change was coming my way. There was no amount of hemming and hawing, no more stalling with the lawyer; it was a done deal. But I came to Cathedral Rock that day as if the visit could somehow slap me into a reality that the rest of the world down here had already grasped. The bumper-sticker credo of the extraction industry proclaimed, "Keep Calm and Frack On." And worry about the future later, if it ever gets here.

From my perch I could spot the fence line that separates the two Starr pastures, east from west. And it was near that fence where I first squeezed the trigger on a living animal. I go back to a November day in 1965, riding in an old Willys jeep, dipping down through a whitebrush thicket and over a small sandy rise. My father is in the passenger seat. I sit in the back, rifle at the ready and eyes on the tree line. Our ranch manager, John Dodgen, is at the wheel. We cruise slowly on. Two does flush from the sound of the engine and dart before us, gray wisps of smoke flashing over the dull Indian grass they part with ease, their white tails wagging behind, propelling them on to the safety of brush and cover.

And then he appears, pausing for an instant between two big mesquites. The midsection is all I see. A snap shot, and the dull thud of a hundred grains of copper and lead on hair and ribs.

Silence. I walk to the dying buck, count thirteen points even before he is dead, look into those kind and gentle deep chestnut brown eyes, and wonder at what I have done.

That rifle had a history. It was a wedding present from my father's best friend to my mother. If anything could be symbolic of how things are down here, it's giving a custom-made deer rifle to your best friend's wife on her wedding day. It is a .250-3000 Savage with a prewar Mauser action. A golden bird's-eye maple stock sets it off as a weapon crafted for a woman, with a polished chromed bolt and indigo-blue steel barrel.

My father's best friend was a funeral director. Daddy's joke is that his old friend Porter "would be the last one to ever let you down." So at a recent service I sauntered up to him and recounted my first deer conquest, thanking him profusely for that great little rifle, that tool of death that presumably transformed a boy into manhood with a simple squeeze of the trigger. He dropped his head at the mention of the rifle and recoiled from me. As he lifted sad gray eyes that had absorbed the sight of a million mournful faces, his voice began to tremble. He said, "I have spent half my life trying to forget that rifle." His words go over me: are we talking about the same thing? "What do you mean?" I stammer. He drops his head and in a whisper places a piece of the past before me, one that I have heard about but never wanted to plumb the sadness of. "That was the rifle that killed Jim Tate."

From what little I know of Jim Tate, I have constructed a man of mythic proportions. I know he must have done something wrong in his short life, but so far no one is willing to say anything about it. And not just because he died an early death, the kind of tragedy that transcends what you think you know about the sadness of the world, or a mournful loss that can never really be fully felt. It only grazes the surface of your capacity to grieve; the bottomless pain you feel has no point of reference in the world you know.

I imagine them, Jim and his new wife of eleven months, driving down from their home in Llano County to a land that was as foreign as it should have been familiar. After all, he was a cowboy, and cowboy country was where he was headed. Jim first went to work here just after high school. Having missed the war by a year, he and Meta Laverne made the hard right turn to matrimony, sealing them forever in the kind and simple reflection of each other, knowing that they thought they had what can never be doubted or destroyed.

To this union came little Sanford Oron Tate Jr. The cowboys at the headquarters used to call him "Little Rusty" because of his shock of red hair, pulling him around the dusty front yard in a makeshift buggy contraption and hitching themselves to it in mock servitude to the toddler tyrant they adored. I can see his mama now: thin cotton dress, hair disheveled by wind, sweat, and the heat of the day. She places him in the cool shade of the front porch hackberry, letting him soak in the wild world that surrounds him.

Elmo Owens was a water well man, a driller by profession, who during a drought was seen as something of a community soothsayer. He had the machinery and know-how to relieve a suffering that knows no bounds. With the turn of a screw and the release of a valve, the cable-tool rig pounded the dirt and sand, fractured bedrock and stone until it caved in and succumbed to time, weight, and the steady thump of Elmo's bit. Elmo provided life-giving water, the coin of the realm in a kingdom ruled by those who had it.

On the afternoon of May 15, 1953, at 4:45 p.m., Elmo Owens headed out the Eagle Pass Highway. He was hauling a load of carrots over to a trucker in Eagle Pass. As he rounded the bend near Peña Street, just as the highway straightens and starts up the hill, he turned the steering wheel of his truck abruptly to the left, directly in front of the Tate family in their car. They had been checking on some steers in a pasture my grandfather had leased over near El Indio. On the driver's side of the gray Ford Crown Victoria, Jim Tate hit the steering wheel with his ribs and sternum. In the passenger seat sat Meta Laverne and Little Rusty.

Jim lived. His wife and child died.

I look for solace where I can find it. Saying they did not suffer seems as hollow as it is trite. The suffering would be borne by the man behind the wheel. I have asked those who were there then. My father was overcome with grief and could not speak of it. A fellow cowboy, Harold Lansford, told me recently it was just too sad to even try and understand. If the aged and infirm from a half a century ago are leaving this rock unturned, why am I compelled to kick it over?

EL, who was there at the time, remembers Jim this way after the tragedy. He would wander the pasture on horseback late at night, staying out till just before dawn, then come in for breakfast at the bunkhouse, avoiding all eyes and casting his head down, his narrow-brimmed Stetson canted forward, armor from inquiring eyes that meant only to help. But there was none. His world was gone. No amount of grief or love or anger turned outward or inward was going to bank a fire that fueled itself on what was undoubtedly the mother lode of guilt. The only thing he had was hatred.

Jim Tate's best friend was R. B. Owens, Elmo's son. He remained his friend until the day Jim died. Such was Jim's character. Your best friend's father has killed your beloved wife and little baby boy. You hold no grudge—you still look him in the eye and call him your friend when you go to town. It is a mystery, and one that gives me hope for humanity.

Jim, wandering the wide-open pastures on horseback with a full South Texas moon to cast a silver light on swaying bluestem, the silhouette of dead mesquite stripped of life but holding fast to an earth that understands it. All alone he rode, with the steady plodding footsteps of his horse, keeping time with a man who wished for what he could not have. Comfort comes but never stays. Its very nature is alien. It tortures by its sordid presence, then retreats to live another day, knowing full well it has a victim at the ready, a feast for the demon that feeds in unrequited time.

E. L. Pond could never be mistaken for another man. He is six foot seven, and his right arm is severed just above the elbow. As a child my conversations with him were at the belt buckle level. He knew Jim Tate well, worked with him, listened to his muffled cry of pain and sorrow in that darkest night. He told me that Jim would lie in his bunk and whimper like some poor sad animal whose only solace came from the sound of its suffering. And if anybody knew about the limits of human suffering it was EL.

In winter 1956, just before Christmas, my father invited one of my Wichita Falls cousins to come to the ranch for a deer hunt. Jim was to be the guide. They had ventured off into the wildest, most remote corner of the ranch, the West River pasture, which is a short distance from the river and Mexico. They found a dense whitebrush thicket and began to rattle, hoping to entice a buck by simulating the sounds of combat, pretending to be mock adversaries in the seasonal quest for a mate. The details are sketchy, but it looks like that .250 Savage discharged when my cousin handed Jim the rifle while crossing a fence, or else it went off when the boy fell forward. The bullet struck Jim in the back, killing him. The death certificate from the coroner lists the injury as an accident and the death as instantaneous. So that beautiful little gun, the funeral director's pride and the one my mother killed her first and only deer with, ended that fine man's life.

◆ ◆ ◆

A smooth sheen of the faintest light floated in from the east as I stood on Cathedral Rock, a shift in hue that prompted birds to take wing. They left behind their raucous chorus—a challenge to their brethren, but a siren's call to me.

That day I didn't have the sound of songbirds for long. In the distance the drone and low, throaty rumble of a 20,000-horse-power Caterpillar diesel engine slowly shattered the morning stillness until it engulfed everything. A multimillion-dollar, sixteen-stage "zipper frack" pulsated 8,000 feet beneath the ranch. The clanging of drill pipe and a sudden burst of fuel to the engine spiked the air. On the horizon, a thick black cloud of carbon puffed into the pale blue sky. Men, money, and machines united to obliterate a silence that has always been one of the best reasons to remain in this remote part of Dimmit County. The dust and the din, the clamor of pipe and the jolting report of a heavy chain the circumference of a tree trunk beating against iron drove me higher on that rock refuge.

In the midst of that cacophony, a scissortail tried to set the daily boundaries of his territory. Arching like a roller coaster between treetops and brush line, he forewarned all who were within the sound of his song that he alone was lord and master of that space. Fanning his long twin-forked tail feathers wide, he slowed his descent and came to rest eight feet above my head. Leaning back, I stared at his slate-gray breast, the faintest trace of pale rose-colored feathers flecking his regalia.

For millennia the flora and fauna of this land have endured a protean climate and finite resources. In the prophetic words of the late and longtime Dimmit County resident Suddy Dullnig, "No country promises you more and delivers less than this place." But when and if moisture arrives, it means mating, and when the inevitable flip side arrives there is no choice but to abandon or die. But now the game has changed. The conventional drilling that had been operating on the ranch since 1923 gave way to fracking in 2008. But because oilmen know humankind so well, they omitted the deleterious effects one would suffer from fracking and began the party with a big bag of money.

It took a while to realize that there was a huge cost. Our precious water was being mined, extracted from our aquifer, and then injected along with a chemical cocktail of toxic carcinogens into the Eagle Ford shale below. This liquid and sand mix explodes and fractures the shale rock. If done correctly and with precision, the frack stays where the engineer wants it to go. But a dip in the formation and an unexpected turn thousands of feet below the surface can lead to disaster and the contamination of your fresh water. At the steering wheel of this juggernaut of a hydrocarbon holiday is a mind-set that urges us to get as much as you can while the getting is good, or, God forbid, before the price of a barrel of crude hiccups and goes from up to down before you know what hit you.

In a pasture closer to the ranch house, I could see my herd of bison grazing placidly, oblivious to the changes around them. Moisture graced the soil, grass sprang forth, and the herd ambled on. They were then and are now the future of this ranch. Without our underground water, we might as well be on the moon. And water is what it is all about: our aquifer recharges at a snail's pace, far more slowly than we are being mined.

Back in 2011, during our most recent one-year drought from hell, I found my way back to church to pray for rain. Just as there are no atheists in foxholes, there are few atheist ranchers, especially when it has not rained in six months. At best I should be described as a wayward and itinerate member of the Holy Trinity congregation in Carrizo Springs. After the service, while visiting with the other parishioners, an out-of-town oilman who was there to get what he could before someone else did casually mentioned to me a fact his company had imparted to him when I brought up the subject of the water used for fracking: "Well, my people tell me that if this boom keeps going the way it is predicted to, Dimmit County only has fourteen years of fresh water left." I nodded, turning my gaze first to the freshly painted beadboard ceiling on that old church and the stained glass window above the altar that depicted the Holy Trinity. Father, Son, and Holy Spirit—I will need them all, for, unchecked, this headlong rush into the chasm of consumption will undoubtedly be remembered as the time when we killed what we had in order to get what we probably didn't need.

I climbed steadily, cautiously toward the summit. Every crevice, every dark hole and indentation could have been a snake den. I slowed my pace, letting my eyes rest on a flat smooth surface just above me. The slab of stone had been rubbed level, creating a vertical tablet in the rock. Deeply incised in the unmistakable cuneiform style of European lettering was one word, faint on the edges, an *R* followed by an *E*. Then mind and logic converged as I tried to ascribe meaning where there may have been none. Leaving one's mark on a rock is something my grandfather would never have done; he left his impression on people instead. He would consider it beneath him and frivolous to take the time and make the effort to carve his name in a natural feature of the land. But somebody did.

I scrambled and searched for more, but time and weather had obscured the message that once must have meant so much to whoever left it. I traced my fingers in the indentations in the stone. I felt more than saw the word: *Recuerdo 1741.* Memory.

Memory, in any language, was the perfect word for that day. I knew I couldn't move forward until I looked back. Like looking in the rearview mirror, I saw the obstacles avoided, the near misses, and the fender benders. The head-on collision could finally be viewed from where I sat.

I stared at the changing landscape below me. What can one man do?

· EIGHT ·

Natural Resources

I paid scant attention to oil and gas operations on the ranch in my formative years, but when the summer of 1980 came and natural gas prices spiked, setting off a drilling frenzy on the ranch, I sat up and took notice. What I remember more than anything was my father's reaction to a well that came in with an initial force and pressure that made everyone hop on the gravy train with glee. He gave an ear-to-ear grin that was reserved for special moments of unfettered joy. It was not long before visions of private airplanes and new trucks clouded all our thoughts. His enthusiasm became infectious, and for the first time I thought, My God, can this really be true? Hydrocarbons flow from a well into a pipeline: your well has come in, and any consequences of its production take a back seat to the benefits of supposedly free money. It was not too different from getting that first rush of adrenaline when I was hunting, following a scent as old as time.

But just like that, before anyone could gain their footing and absorb the impact, there were hundreds of little pump jacks, tanks, separators, and diesel compressors all working in a glorious harmonic convergence to extract and produce something that nature had kept secret since way before our time began.

We had a big wide-open ranch with plenty of everything.

What difference did it make to sink a few wells into the pasture and pump up what we were lucky enough to be sitting atop? Unconventional drilling was still a ways off. Until then, the oil dribbled more than flowed. But then the pressure dropped. Boom and bust, ebb and flow.

When I was twenty-one years old and had dropped out of college for the second time, I sat next to my father at the Number 5 pens on the ranch. On a timeworn, rundown joint of drill pipe that served as the top rail for the alley that led to the squeeze shoot, I turned various matters over in my mind. One kept resurfacing: my future.

As we watched, the cowboys bulldogged the young bull calves. The calves bawled and bayed while a six-inch Case pocketknife was passed briefly over the orange flame of a Zippo lighter and then neatly pared the skin of their scrotums. The testicles were then squeezed out, severed, and tossed unceremoniously into a wide chipped porcelain dish to be fried up and served at dinner, *hors d'oeuvre a l'enfant sauvage.*

What I had at that moment was my father's attention. Because of my itinerant state as a would-be scholar who did not want to take tests or sit in a classroom, I asked him what career path he thought I should wander down. The question missed its mark as he fumbled for his fatherly footing. He gazed across the alleyway into the pen where the mother cows were pacing back and forth, watching their young sons graduate from rambunctious, testosterone-infused juveniles to placid dolts who were now destined for a trough of corn in the feed lot and a journey down another alleyway, at the end of which stood a man in a white smock. After a pause that seemed interminable, my father answered, "Well, I guess you could move up to the ranch at Llano and breed Hereford bulls." That answer seemed infinitely better than watching what was going on right in front of me, but it was still not in tune with who or what I thought I might be becoming. I honestly didn't know what to do next. It seemed like time was speeding up.

Before I knew it, I fell in love. I married. I had my own family. Finally, I was home.

◆ ◆ ◆

The liberation of a lifetime came wandering into my life in the form of the largest terrestrial mammal on the North American continent. People always ask me, "How is it that you chose buffalo to raise?" I often respond, "I didn't. They chose me." They are here to stay, and from the moment they arrived they have influenced everything I do.

I once found a map by a French cartographer named de Beauvilliers. It was dated 1720 and showed where he thought the Rio Grande might be and the other significant rivers in Texas and Mexico. In bold type just above where the ranch stands today was written *Pays de Boeuf Sauvage*, the country of the "wild cow," or bison. It was a tattered old remnant of one man's imagination, and it was the color of coffee-stained paper. Sea monsters arching half submerged and raging wide mouthed and vengeful were in the Gulf of Mexico. Missions were drawn like tiny little houses with a cross to mark Presidio del Norte, and diminutive hills marked the Rocky Mountains. It was a map more of the imagination and the emotions drawn more with the excitement of the moment than with exactitude and precision. But the bison would not be denied its place in a landscape where it was the dominant mammal.

They had been long gone, but I bought a herd. Wide-eyed and wonderful, they thundered off the double-stacked cattle truck, bucking and twisting after nineteen hours in an eighteen-wheeler, hauled in from the High Plains. They had roared down the interstate to the ranch. Moms and dads in minivans poked bored children in the ribs and banged on the window: "Look, there goes some buffalo! I didn't think there were any left!"

I first heard of the bison in Dimmit County from the book *The Explorer's Texas*. My brother gave it to me before the family troubles began. To me it is almost like a fairy-tale book come true, describing the land and waters of my state as I wish they had remained: grasslands and open parks where animals wandered to their hearts' content, raptors doing what they did best with precision behind the eye and grace on their wingtips, a world where everything that happens is as it was meant to be.

The book gave 1781 as the last recorded sighting of a wild bison in what was to become Dimmit County. They were here by the tens of thousands, passing through, grazing the seacoast bluestem down, and then passing on to other pastures. When you fly over the land in a small aircraft, you can still see the wallows where the thundering herd would fall to the ground and swing their great bodies with force, rocking rhythmically as they rolled on the dirt.

My work became my life, and my life became my work. In the blink of an eye I populated the ranch with the largest land mammal on the North American continent. The buffalo is the oldest continuously harvested herbivore in the country, and the sight of a large herd of these magnificent creatures roaming a land their species had not graced for two centuries thrilled me to my core.

One week after I bought 250 head of bison my neighbor, sixty-one-year-old J. R. Hamilton, spotted me and my nascent herd in the West Annex pasture. It had been a good year for grass. Tall native bluestem and tanglehead grass laid a thick mat of green; it is the most bison-friendly land on the ranch.

JR, never at a loss for words, hollered over the fence to me as I sauntered toward him; the herd of bison placidly grazed behind me. All five feet six inches of him were lit up with glee at the sight of what had been missing from Dimmit County for more than two hundred years.

"Sunny, look what you got, man. They're going to do great down here!"

Since the Pleistocene these bison have mastered the art of endurance. They scratch and rub on the mesquite, that other survivor down here, and don't do anything they don't have to. That's their secret. They live a measured existence, and as long as they have grass, water, and each other, they thrive. Amazingly, the roar of F-16 fighter jets overhead doesn't seem to bother them. That might be seen as a good thing, but during the great slaughter of the herds during the nineteenth century it was precisely that predilection for tranquility that was almost their undoing. It was then that they were shot en masse by men leaning out the windows of trains, for sport, or to decimate the food source of the native peoples and force them to surrender. The Great Plains was strewn with skeletons bleaching in the sun after the big herds were gone, the remains of some 30 million herbivores. By 1900 fewer than one thousand remained.

Conservation, if it happens, tends to come at the eleventh hour. Some of the bison were spared and shipped to the Bronx Zoo or were bred by private ranchers like Charles Goodnight, whose wife Mary Ann insisted that he rescue any animals he could. The survivors of the Goodnight herd have increased in number to more than one hundred animals, and I am lucky enough to have some of their progeny at the ranch today. There is a special dignity and grace in their movements.

Some may see raising bison as my folly here in the *despoblado*, the no-man's-land. But I have always believed in sustainability, in keeping the numbers in line with the grazing capacity of the ranch. At the same time, I want to provide an economic benefit for myself and my family and the individuals I employ.

I started with a visionary's pride. My first mentor was a member of the Coahuiltecan Nation. Ted Herrera has always been a close adviser and friend whenever I had questions about the well-being of the bison and my relationship to them. He is a respected elder and has studied the history of his people and their relationship to the bison in South Texas. At the very root of my venture was the notion that the animal is sacred to so many people. This was going to be holy meat from holy animals raised in a holy way.

What I did not count on was the unholy drop in live animal prices. When I bought my foundation herd of bison in 1999, my business plan was to run a traditional cow/calf operation. Calves were selling at an all-time high, and this seemed to be a logical and profitable direction for the ranch. Within a year of my purchase the live animal price had fallen by 80 percent, and we were rapidly descending into a drought and its deadly consequences. I was overstocked, there was no rain in sight, and the market price was nonexistent. You could not give a bison away.

I had two choices: try to liquidate the herd by selling the bison at a pittance and going out of the business before I ever got started, or start harvesting the bison and establish myself in the grass-fed/field-harvested meat business. I chose the latter.

We are a grass-fed-only operation that depends entirely on rainfall to sustain the native grasses and forbs the bison eat. One of the great botanical paradoxes of South Texas is the mesquite tree. As it sucks moisture from the land it produces a bean that is up to 13 percent in protein and as high as 3 percent in fat. In drought years the tree believes it is going to die and produces bumper crops of beans, which the bison then eat, providing them an enormous nutritional boost.

A combination of understocking, the natural grazing habits of the bison, and drought emergency measures are what we rely on. Reducing the number of animals to take the grazing pressure off the ranch was vitally important, and we did this by creating the first bison field-harvesting operation in Texas that is under the direction of the Texas Department of Health.

The next step was to seek out markets. Texas was and is cattle country, and the beginning of my bison venture began in the days before the local, grass-fed, and sustainable consumer had made their impact on the market place. I had to put my faith in education and history. If these magnificent animals had once sustained the native population, surely the public would respond sooner or later.

In 2001 I began to sell my product at Republic Square Farmer's Market in Austin and then expanded to the Pearl in San Antonio. Restaurants soon caught on, and by 2004 the demand was exceeding the supply. Rains returned and the work paid off.

What separates a nonconfined, genuinely grass-fed and field-harvested bison from the rest of the pack is cleanliness, the purity of its protein. When I brought home the first animal that I had taken correctly, I defrosted the ground meat and set it on the kitchen table. My wife, curious, inspected it, and as our DNA has programmed us to do, she sniffed it, then cocked her head to one side and said, "It smells clean." No more potent phrase could be used to describe the unique characteristics of what was offered on the table before us. I don't think I've ever heard such an attribute concerning a feedlot steak or a chicken that has never seen a sunrise.

We have been in and out of numerous droughts that have challenged our business, and overgrazing is still my biggest nightmare. In 2005 I fenced off and planted an irrigated field where I could grow improved varieties of grass that would serve as grazing buffers during dry spells. Later, I built a unique cross fence that divides the ranch in two. This fence allows wildlife such as whitetail deer, bobcats, and javelina to cross under the fence at designated places that are just slightly elevated and reinforced with pipe braces. This way I can rotate the grazing while allowing the wildlife the freedom to roam the entire ranch.

Our continuing goal at the ranch is to improve the health of the land and the lives of those who are intimately associated with the ranch through the production of bison and the management of our wildlife. Biomimicry and using the land as nature intended is the only way a sustainable ranching operation has a chance in this part of the world.

We operate with one full-time employee who, with the help of various family members, is responsible for the vast majority of day-to-day operations. Alfredo Longoria and his extended family of brothers, uncles, and nephews are all natives of Dimmit County. On many occasions the Longoria family makes the ranch available to the teenagers in the community for hunting and horseback riding.

In order to diversify, we've been keeping between ten and fifteen active bee colonies on the ranch to produce honey. In April and May the bees gather the nectar from the native trees, from the guajillo, honey mesquite, catclaw acacia, paloverde, and guayacan. The result is an extremely delicate and flavorful desert honey.

We rely on the bees as an indicator species for the overall health of the land. In addition, we believe that a healthy predator/prey relationship is vital to the well-being of the land. Healthy coyote and mountain lion numbers hold deer populations in check. Rattlesnakes have their numbers thinned by roadrunners and bison.

When it comes time to kill a bison, my ranch foreman, Freddy Longoria, and I go out the day before a hunt and look for tracks. Bison have a practice of establishing routes through the pasture and then sticking to them, so usually they can be found without too much trouble. Once a herd is located, we put down some range cubes and let the animals settle down so we can take a look at which ones might be suitable for killing.

We start hunting at daybreak. Early morning is a quiet time, when stillness and sunrise merge to give foundation to the day. Dawn has always been a time to collect my thoughts. I need to think about the fact that the lives of living creatures will end by my hand. It's strange to say, but conscious killing gives me a sense of release. It's an intimate act, a primordial connection that happens when you get that close to life and death. But there are also times when the fine line between them is hardly visible.

◆ ◆ ◆

While my mother was still alive, things were cordial between my siblings and me. We all had children by then, and while there were occasional disagreements over things like who got what weekend at the ranch there was little of any substance to pull us apart. But when my mother descended into the late stages of cancer in 1993, a lot of things began to change.

I drove out to the hospital to see her for what turned out to be the last time. I was remembering the important things she tried to impart to me when I was a small boy. In the library of our home in San Antonio she once found me staring at a book of Greek mythology. Then she took my left hand, touched the ring finger, and told me that the ancient Greeks believed that there was an artery that ran directly from this finger to a person's heart. She taught by myth and history. In my bedroom at home she hung a small print of Scotland's warrior king Robert the Bruce, a man who had known the humiliation of defeat but was determined to rise above his adversaries. The story goes that he had lost two significant battles to the English and was considering surrender. Taking refuge in a barn, he collapsed and then awakened to the sight of a spider that was attempting to swing from a rafter back home to its web. The first two attempts came up short, but on the third the spider landed securely on its mark. The warrior king took this as a sign that if he did battle once more he would prevail. He did.

I knew when I entered my mother's hospital room that her battle was almost over. Her skin had taken on the translucent look of very fine parchment. The chemotherapy had long since done its work, and she was so weak she could not lift her head when we spoke. Her husband, Bill Negley, was there. He stood at her bedside looking down at her, and then pointed to the floor, where my mother's delicate little slippers were lined up next to the bed. He turned his back and stared out the window, his broad shoulders heaving.

Bill left the room to go home. It was over, and he knew it. I was alone with my mother in that hospital room.

As I sat with her, I clearly remembered the time when I was seven and she took me for our one and only mother-son fishing expedition on the San Pedro Creek. From the first tentative tremble of that orange and green cork dancing before us in the water to the rush of witnessing it submerge, we laughed together as the catfish desperately realized he was hooked and began to fight against the rod. I extracted a living creature from the depths of a dark green pool and presented it to her. What stays with me from that early spring morning was her pride in and astonishment at my ability to catch that fish. As I remember, the fish could not have been more than ten inches in length, but you would have thought by my mother's reaction that I had landed a denizen of the deep the size of Moby Dick. She was beautiful by that stream bank, and as I slid that catfish on my stringer I felt both empowered and important in her eyes.

I may have been initiated into manhood by men like my father and EL, but it was my mother who taught me about beauty, who encouraged the part of myself that wanted to know the names of birds and flowers. I wished I had been closer to her. Under those hospital fluorescents I hardly recognized her. "Sunny," she whispered. I leaned closer to hear her.

"I want to tell you something. You all are going to have trouble."

Those were the last words she spoke to me before she died, not words I had expected to hear. What could she mean? She glided off as a steady drip of morphine coursed through what was left of her body to bring her peace at last. I couldn't ask her what she meant, what I was supposed to do.

On the night of her death, as I was driving home, the news came over the radio that the writer, professor, and environmental activist Wallace Stegner had also just died. A light mist of rain enveloped my car as I crept down the road at a snail's pace, lost in a "between the worlds" moment. He had become confused a few weeks earlier while driving in Santa Fe, New Mexico, and had run a stop sign. Hearing that news and having just witnessed the death of my mother, I sensed a connection, for Stegner was not only one of my heroes but also had a close and intimate association with my mother's aunt. Elizabeth Simpson—Aunt Ibby, as she was known—was the wife of Claude Simpson, who for many years was a teaching associate of Stegner's. Stegner represented standing up for what is right and speaking one's mind, and that night I think I started to consciously try to do both for the first time. It was as if I had stepped into some eternal cosmic convergence of life, death, and the never-ending quest for the answer to "Why am I here and what am I here for?" Wallace Stegner felt like an ally.

My mother's death and his set me on a path apart. In ways I

didn't understand at the time, I was being separated from the herd.

•   •   •

Signs of struggle for territory have always been visible on this ranch: Spear points. Knives. Artifacts that spell out forcefully and in no uncertain terms just how deadly ancient warfare can be. But there are relics from more modern times, like .50-caliber shells from World War II, when the land was used as a gunnery range.

Over the fireplace mantel in the kitchen at the big house are two firearms set into cement—an unusual and unnecessary method of affixing that seems bizarre, but the guns were not always in cement. My father told me that when the commanding officer of the gunnery range left the headquarters, he decided to take a little souvenir with him, so he took the antique guns. When he got the ranch back, my grandfather marched up to the man's home in San Antonio, walked into his living room, and asked to have them back. He then returned to the ranch and had the rifle and pistol cemented into the wall above the fireplace.

On the southern boundary of the ranch there is a relatively recent scar, one that is wide and white, a glaringly stark line that lies in contrast to the green buffelgrass and soft ash-gray leaf of the cenizo that lines the border of our boundaries. A seven-mile caliche road was bladed through the ranch when it was divided between my father, my brother and sister, and me.

It's an old story in Texas ranching history. Mineral rights in this state have created many legal entanglements as families sort out their holdings and put up fences, legal and literal. There is

nothing that can create more division within a family than trying to balance individual desires with land use. The visions of
the original homesteader often are not shared by generations to
come. But the irony isn't lost on any of us: the money that my
grandfather used to buy this ranch came from his stock and his
hard work when he was a driller for Texaco.

✦   ✦   ✦

A bend in the river is called an oxbow, a languid passageway that
water cuts when it courses a new path, slow and steady off the
main channel, until it separates itself and creates a life apart.

This land is where I chose to take my stand. I had a fight
on my hands. A younger brother and older sister had our dying
mother sign a legal document that transferred some of my inter-
est to a trust that was controlled by my brother. In a so-called
deathbed transaction, what was mine came under their control,
and that was that.

After one of my mother's debilitating rounds of chemo, I went
home to see her. She was lying on a narrow twin bed. The cur-
tains were drawn, but there was enough sunlight to illuminate a
face ravaged by the poison meant as medicine. She tried to explain why she was putting my younger brother in charge of the
majority of her estate. Evidently his legal training swayed her.
Then she whispered words that sliced both ways.

"Well," she said, "at least *you're* honest."

Three years later, in 1996, my father and I decided that we
wanted out of this arrangement, which had already caused us
so much unhappiness and discord. My father wanted to sell a
portion of his land, so the great divide became an even more

chaotic choosing up of sides. My father aligned with me. Joseph, my brother, was running the ranch through my mother's estate, controlling her financial assets through the labyrinth of trusts and partnerships set up by her lawyers.

The trick here is to make every thing so bloody complicated and confusing that you sign up and before you know it you are standing in the middle of the street with your pockets turned inside out. Were it not for my lawyer, I would not have a pot to piss in. Daddy needed me in the fight, at least for a time. And then one day our paths diverged.

That day came when I found out he and Joseph had decided to lease the deep rights on the ranch for drilling and exploration. Behind my back, without telling me what the plan was, my precious land that I had fought so hard for was being subsumed by the stroke of a pen.

I found out about the lease from one of the pumpers in the field. Funny how you remember exactly where you were, who the person was, and the look on their face when you tell them they are telling you something you should by all accounts have known yourself a long time ago.

So I beat a path to Austin, called my lawyer, kicked and screamed, and then accepted where I was. I fell back on some old patterns that I am not proud of. Victimization is a deadly syndrome and a pernicious frame of mind that has suffused so much of my life as it relates to my family. To guard against it to raise my awareness of how I can break the learned behavior that has done nothing for me but a great deal to me.

It was a cold clear night in mid-January. My line in the sand was already drawn with Daddy. I told him that he had to save my land, at least the part that had not been drilled on—the only land left out of my grandfather's original 65,000 acres that had not been defiled by the oil companies and all they bring with them. I told him either you save my land or off you go. It was that simple. I went to bed nervous and agitated, knowing full well that my father was not going to save one square inch of the ranch from the drill bit if it meant giving up a nickel. Euripides could not have written a more poignant final act to a father-son tragedy. He arrived with bowed head and shuffled up to the kitchen porch, then walked through the door, and we sat down to seal our respective fates. I did what I could, but I knew it was over.

◆   ◆   ◆

In some ways, our family has since made some peace. There are fences between our properties, though, dividing a place that was once whole. The fences are a reminder not only of our differences but also that we share a split estate. Landmarks like hilltops, trees and water troughs are physical places where you loved or cried or killed, and the land itself is like a living organism that took up residence inside you the day you first set foot on it. You feel that to lose a certain stand of bluestem or an oak tree would be no different than cutting off an arm.

If I have one hope, it's that my own children will never have disputes over the land that will one day belong to them.

A few years ago my wife, Sarah, and I were taking one of our walks together down by the creek. My ranch is next door now; it had been a long time since I sauntered by that beautiful stream. But on that walk we moved slowly up the old Camino Real toward the big house, the old headquarters where my brother and his family live. Before me, half-buried in the sand, was the tiniest little arrowhead, called a bird point although it could also be used to dispatch larger animals. It had a serrated edge. The base was intact. The only damage on the projectile was an almost imperceptible chip on the very point. It would be amazing if we found that tiny little chip lying on the Camino Real, where it had been apart from its kin for more than four hundred years. Santa Anna rolled over it on the way to the Alamo, Zebulon Pike on his trek across the continent.

Sarah found that chip, lying several feet from the arrowhead.

My family was splitting up the ranch at that time. Finding that arrowhead with Sarah and reuniting it with itself seemed symbolic of our life together. I had felt broken and saddened by the ruptures in my family of origin. Finding that arrowhead held out the promise of healing.

I named my portion of the ranch Shape Ranch. The shape of things to come for a growing new family: Sarah, Hugh, Asa, Patrick, and Evelyn.

I decided to bring back the animal that had once freely roamed the land for ten thousand years and had been hunted nearly to extinction. The past and the present converged. X marks the spot.

NINE

The Arc of Instability

You know it's spring in South Texas when you drive down the highway and even with the windows rolled up and A/C on you're engulfed by the sweet smell of our most pungent acacia. In Spanish, blackbrush is known as *chaparro prieto*. This shrub is impenetrable—it can grow to a height of more than ten feet. It provides one of the most distinctive aromatic experiences that a person can have in the brush country, a scent so overwhelmingly intoxicating that it tends to make you overlook the thorns and spines that make it such an effective source of refuge.

That dense cover provides shelter and habitat for all manner of mammals, reptiles, insects, and birds. Long pale-yellow catkins suspend themselves like miniature Christmas decorations and waft in the breeze. These blooms are a major source of pollen for my bees and butterflies.

But there is another blackbrush down here whose odor is anything but sweet. BlackBrush Oil and Gas acquired a lease on my ranch without my knowledge in one more back-alley family maneuver, one that took place behind closed doors. My mother's estate, which included the ranch, was known as a split estate. My brother, a lawyer and one of the estate's executors, had made the decision for all of us. In a split estate, mineral rights are primary, surface rights secondary. I didn't find out about the lease being signed until the day after my oldest son graduated from college. We were at the Phoenix airport changing planes to come home when I got the call from my lawyer that the lease had been signed. I was blindsided and had no recourse but to acquiesce. As I was discussing all this with him by cell phone, I happened to see an old friend walking through the terminal. By strange coincidence, he was a major investor in BlackBrush. I approached him and said, "I hear that a company you've invested in is going to be drilling on the ranch." He said, with a grin on his face, "Yes. It should be a great deal for you." Except I didn't want them to drill and told him so. A quizzical look and furrowed brow creased his face. *"You don't?"*

When we parted, I knew our friendship was strained. He has always considered himself a friend of nature and believed the oil and gas producers would be courteous and clean up after themselves. He has since sold his ranch and moved to a higher elevation in West Texas, to a place where *he* controls the mineral rights.

♦ ♦ ♦

My home sits directly downwind from a BlackBrush well located on my brother and sister's property. A thick column of black smoke is periodically emitted from its flare stack, a sickening hydrocarbon belching that sends volatile organic compounds wafting into the sky and drifting toward my home. When the wind shifts to the south it drifts toward Freddy's home. Benzene, toluene, ethylene, and xylene have all been detected there by Texas A&M researchers and the Texas Commission on Environmental Quality (TCEQ).

Since 2014 I have been barking at the door about this issue with TCEQ, the operator (which is now Endeavor Natural Gas), and the trust that administers our mineral estate. I have a three-ring binder that is four inches thick with documentation of what are known as fugitive emissions at these sites. They have caused nosebleeds, asthma, vomiting, and temporary blindness to my employees and their children. My maladies are not mine alone; this is an issue all over the state. I met with Texas speaker of the house Joe Straus and voiced my concern about the lack of enforcement. He was visibly shaken by what our group had to say but confessed that there was little he could do. The lack of response from the state is unconscionable, and the physical and emotional fallout are intolerable. Like just about every other mystery in Texas, follow the money, for those emissions flow up to Austin and Washington, morph into greenbacks, and fill the pockets of our administrators and elected officials. Meanwhile the people who elected them are poisoned.

This danger has come from a great human passion that has been unleashed on our land, an avarice resulting in an everlasting revelation that is unforgettably rooted in the effects it has wrought. The irony is that before he died, George Mitchell, the father of modern-day fracking, prophesied that what he had pioneered would be misused by those who played fast and loose to get more at any cost. Just as J. Robert Oppenheimer and Albert Einstein regretted their role in the development of the atom bomb, Mitchell knew the attitude of a certain class of oil and gas operator, corporate scallywags who will always tell you what you want to hear, then turn a deaf ear when the truth comes raining down and they hide behind the billowing skirts of the Texas legislature.

♦ ♦ ♦

I would not have the Shape Ranch today were it not for the extraction of hydrocarbons. Ranching is a marginal economic endeavor at best. High labor cost and variable commodity prices conspire all too often to deflate a rancher's best-laid plans. And then there's the uncertainty of rain. Even though I was raising bison sustainably and selling the meat as well as leasing the land at a premium for deer and quail hunting, I was barely squeaking by. Royalties gave me resources to invest in the ranch and the well-being of all its inhabitants, animal and human, and to diversify and flourish.

Royalties or not, it was devastating to find out that unconventional drilling would soon begin on my portion of the ranch. It was even more devastating to learn that the lease had been signed three years before, without my knowledge.

In 2011 Dimmit County got its first taste of both the greenback intoxicant and the accompanying environmental hangover of the morning after. A veritable firehose of money was turned on for landowners who were being paid for leases, for frack water, and for the damages caused by the operators. When you pulled up at the county courthouse to vote or go see your commissioner, you had to park three blocks away because land men had taken all the parking spaces. They swarmed around the pretty young county clerk's assistant, who was pulling reams of files and running a copy machine at warp speed. The first step in the process of becoming an unconventional extraction colony had begun.

More than four hundred trucks a day raced down our little one-lane county road to drill that first well. White clouds of caliche dust rose up, powdered so fine you could not see out the windshield.

It was like driving through a New England whiteout blizzard, only the temperature on the dashboard of my truck read 104 degrees. The road got so potholed and washboarded that when the school bus came down the road one afternoon with a load of children, its steering wheel came off in the driver's hands. The bus missed an eighteen-wheeler full of fracking fluid by just a few inches.

Driving down the Dentonio Road, straining for a clear view of the road ahead as I inched along enveloped in a cloud of dust left by a truck full of fracking sand, I caught a glimpse of the tattered remains of a dime-store American flag flying on a neighbor's gate. It resolutely waved at every passing welder and roughneck as they sped with unbridled enthusiasm toward a future that couldn't be sustained.

Another well soon followed. It was accompanied by heaps of garbage on the side of the road, lost truckers who wandered all over the ranch, and an eighteen-wheeler jackknifing and discharging its load of contaminated frack water into the pasture. The third well brought the issue to a head. They had to run a pipe from my two strongest water wells to the frack water pond, and my well could not keep up with what they needed. They were pumping away. All was going as planned until they drained the tank I was pumping into right down to the muddy bottom. When I saw this I asked them to stop for a few days so my tank could recover and my bison would have the water they needed. In addition to not having water to drink, the animals could easily wade into the mud for that last sip of water and get stuck. A slow and painful death was sure to follow.

The operator interpreted my asking them to suspend pumping for a few days as a refusal on my part to provide any more water to them. That was not my intention at all. Their response was to not frack the next stage. Not one peep from them, like "Hey, Mr. Sunny, how about turning the water back on?" Instead of offering to truck in a few eighteen-wheelers to refill my tank so the bison could have a drink, they just folded their tent. Their inability to communicate with me cost them somewhere in the neighborhood of $250,000. But the rush was on, so they raced away to the next. There was more where that came from.

Human nature, as far as I can tell, does not lend itself to either moderation or reasonable behavior when it comes to the extraction of anything. We pump, mine, and dig until it's all gone. I can't help but make the comparison of the near-extinction of bison with the unbridled extraction of minerals by unconventional drilling. In both cases, heedless greed is the motivator. What seemed abundant isn't. Worse, we don't know something's going until it's already gone.

In our little corner of the state of Texas, water governs life and one's enthusiasm for it. My grandfather's favorite saying when asked how much rain the ranch got every year was this: "We get around twenty-one inches of rain a year. That usually comes in two ten-inch rains, and we never get the other inch." That was all too true. So when rain does come, and it does eventually come, we are known to let elation take over.

I built a water tank when I first got my portion of the ranch in 1999. I shot the elevations and dug a test core to ensure a deep, solid impoundment that would hold water during drought. Two weeks after completion, Hurricane José blew in from the Gulf of Mexico and unloaded six and a half inches of rain in just nine hours. The tank filled up overnight and backed water into an ancient stand of mesquites covering four acres. Flights of pin-tails and mallards arrived with the first polar front. Setting their wings for descent, they glided effortlessly between the tall trees and landed in a perfectly controlled stall. Paddling in unison, they tipped down like fat feathered teakettles as they fed on water bugs and minnows. Water brought life to life.

But 2011 turned out to be the year it didn't rain. A miserly three and a half inches fell that year, and ninety days straight of temperatures over 100 degrees put the icing on a grim cake. Des-iccation began slowly, then with each passing day plants, animals, and anything and everything, including the cactus, shriveled up. I tried to turn my full attention to my bison, but the surround-ing physics of fracking surmounted and suppressed what the San Francisco windmill could never overcome. On one visit to the mill I peered over the side of the pila to see a drowned Harris's hawk and a seven-foot-long blue indigo snake, both dead because they were lured to something they needed and were trapped when the water level fell.

The San Francisco gave up its last drop of water. The water had been legally drained and depleted by the extraction of groundwa-ter just to my north through an archaic law called the "rule of capture" that defies commonsense. A well field on my neighbor's ranch pumped thousands of gallons of water per minute and then injected that water into what is known as a slick frack a mile and a half below. It disappeared forever. Another man's straw had drained my glass.

I shifted my focus to the deeper remaining wells and began a metering program to help me monitor my consumption. The San Francisco's death was a warning shot across the bow.

The heat and the drought continued. It stopped raining. Shade was nothing more than a more comfortable place to die. At the shallow edge of a whitebrush thicket I found the scattered remains of a young fawn whose bleached bones lay in mute testimony to a hell on earth.

◆ ◆ ◆

I relive that time over and over again as if I don't know how it ends. I'm driving through the dust, and the life has been beaten out of every living thing. When I'm a good three hundred yards from the Asa tank, I park the truck. Despite the dryness there are still signs of life in the road. Overnight the delicate and almost imperceptible imprints of a dung beetle have been pressed ever so gently into the grains of soft sand, a tiny reminder that life endures during times of trouble. The sideways curve of a kingsnake has thrown a small bunker of sand on either side of its path, a cold-blooded reptile in hot pursuit of a wood rat. The rat's demise is marked by divots and mounded soil piled by the final involuntary thrusts of its flailing hind legs before it was swallowed whole.

From their lofty perch, a pair of crested caracaras watch me as I quietly slip forward. Their twin talents of observation and patience are the keys to predator survival during time of drought.

Just shy of the dam I pause. The unmistakable smell of death assails my nostrils, that sweet and fetid mixture of a life that has passed and now gives off the sad odor of reclamation. With every deliberate, crunching step the aroma gains strength, and as I crest the tank dam my fear is justified. I can't help but stare with morbid fascination as a coyote curls back its lips and thrusts polished canines into the bloated remains of a yearling. The mature male coyote pushes his mouth forward, convulsing neck and shoulder muscles, spastically hitting the poor buffalo's bloated body like it was a furry punching bag. The jaws sink deeper with every bite, and as the hind legs of the little heifer convulse the coyote extracts the soft downy wool between its hind legs. The coyote spits the chocolate-colored tufts into the hot dry air. It momentarily pauses to watch the tufts float away, cocking its head to the side as they drift skyward before being caught by a hot downdraft and deposited on the rippling surface of what little water remains.

The heifer's sphincter expands, revealing the first bloody pool of the dark crimson treasure to come, even as the drought erases the last twelve years of toil and the joy of returning a once-native creature to its rightful place.

The turkey vultures arrive. On long black wings they soar in slow arcing circles, rising and falling as they telegraph the latest news bulletin to their brethren. In a steadily building progression the vultures' numbers mount. There's a hierarchy, and at least twenty begin their descent, hovering patiently to watch the one that got there first.

From the corner of my eye I catch the slightest movement. Behind a mesquite at the edge of the pond stands the heifer's mother, a fully mature bison cow gaunt and drawn, mouth and nostrils caked with the telltale slimy green remnants of a cactus. The cow's eyes are glassy, bulbous brown orbs calmly witnessing the devouring of her calf. The cow then turns and walks away.

There wasn't a hole deep enough for me to crawl into.

Everybody has a last straw. That was mine. I had a choice: fish or cut bait. I threw my hook in and started to chum the waters, holding on and setting the drag for the fight of my life.

TEN

Politics

Most days I blow through Carrizo as fast as I can to get to the ranch, but one day in May 2011 I stopped at the Wintergarden Groundwater Conservation District. It was the day of the monthly board meeting, and I came as a guest, not to sit there politely and listen but to put an item on the agenda.

The district's hydrologist, a learned and thoughtful individual as well as a great teacher, was presenting to the board. My proposal was to have this same presentation done at a later date for the whole community.

Just as my agenda item was set to be opened for discussion and a vote, one of the board members, a lawyer, said he had to leave to meet a client. Suddenly there was no quorum and no vote. I was disappointed but left it at that. On the ride home the wheels in my head started turning as fast as those on my truck, and I began to feel like I had been sandbagged. No action on that agenda item meant no information to the public, which meant business could continue unimpeded. Had I just borne witness to the most damnable of all human traits, information avoidance and the convenient belief that what people don't know can't hurt them?

These are the circumstances that drove me to climb Cathedral Rock to get perspective in the spring of 2012. The sense of helplessness and fear had been overwhelming. It was time for me to stop complaining and start campaigning.

I became a candidate for a seat on the board of the Wintergarden Groundwater Conservation District. My strategy, if you can call it that, was to motivate the Hispanic community to vote for me because they are the county's predominant voting bloc. I went to the political and social godfather of the county, a man who knew the community better than anyone. He told me whom I should contact and where I should be campaigning. Let's just say it was an unconventional approach. What's even stranger is that my opponent, the lawyer who had to leave that meeting, was my lawyer. No hard feelings. We both go to the same church, too.

My strategy paid off, and I beat him by thirty-seven votes for the open seat on the water board. I was off and running.

I began my term as a board member by just listening and reading about water law, hydrology, and recharge, and in general trying to educate myself on a subject that I knew was critical but that I had also taken for granted. The boom had graced us with drawdown and injection well conundrums from which we couldn't escape. The first and most obvious was the drawdown in our aquifer. The hard number was that one-third to one-half of our annual recharge every year was now consumed by fracking.

The Eagle Ford was the third most rapidly developed oil and gas field in the world. More than 15,000 Eagle Ford wells have been drilled since the first discovery well in La Salle County in 2008. The drilling companies have used more than 28 billion gallons of fresh water in Dimmit and La Salle Counties to fracture the formation. The sand, most of which is transported all the way from Wisconsin, is used to expand and keep open the fractures in the shale. The volume used so far could fill the Houston Astrodome six times over. It has produced enough gasoline after refinement to fuel every licensed driver in Texas for four years.

Despite the riches from mineral wealth underground, we struggled as something vital to our survival began to disappear. No matter what you do on top of the land or beneath it, it takes water.

When an oil well is spudded (a term that makes the process seems almost bucolic), water is employed to lubricate the drill bit as it rotates through the various formations. Once they are past the water-bearing strata, the drillers switch to a diesel-based fluid for the rest of the drilling. On balance these operations use a minimal amount of fresh water, but when it's time to frack, they'd better have the water in place or it's all over. Conventional drilling requires much less water, but without the fracking you don't have a chance in shale, and to frack you must have a copious and readily accessible water supply.

When I asked my operator's water guy how he planned to get it, he answered with candor. "I hunt water wherever I can find it. I hunt it down." That's right, I thought to myself. He hunts down water and he kills it. In our area of the Eagle Ford 80 percent of the frack water is lost in the formation, while the other 20 percent is injected into disposal wells with the hope that the fluid will stay put for the rest of life on earth. He is hunting down something that should be on the endangered species list.

Ranchers in my neighborhood, myself included, had the good fortune to have a relatively clean and accessible underground water source. But it was time to ask: Is it worth risking that resource for the finite carbon entombed below? Yes, you say, if you own your mineral rights and can sell that precious water for one dollar per barrel to an oil company. But if you are a landowner with no such rights whose water is being drained by a neighbor's well that's pumping water day and night, taking in some cases three months to fill a frack pond, you're out of luck.

◆ ◆ ◆

I drove west from I-35 to Carrizo and passed a dozen or more large white plastic banners flimsily attached to barbed wire fences. Printed in bold black letters were the words "Fresh Water for Sale." When I looked down under the deceivingly green canopy of mesquites that lined the sides of the highway I saw two things: prickly pear cactus and red dirt. The plant life was either gone or going dormant to survive. Drought and the devastation it brought prevailed, with only isolated patches of green where irrigation systems using groundwater tried in vain to make up for the missing rain. According to the investment advisers at Ceres, an investment organization that tracks and recommends companies with sustainable water management practices in their energy quest, Dimmit County stands alone in one rather shocking and troublesome category: we have had more water extracted from our aquifer for fracking than any other county in the entire United States. But is it any wonder that there is so little regulation and enforcement of pumping-related groundwater usage? Why worry about a three-hundred-foot drop in the aquifer if the shale play is generating $60 billion worth of economic impact? Only Alaska's North Slope and the gigantic Ghawar Field in Saudi Arabia reached a billion barrels before the Eagle Ford. In this era we are subject to a tyranny of the urgent.

What it really comes down to is speed. The Carrizo sandstone at the base of the old San Francisco windmill tells the story of water movement through rock. It is slow, painfully slow. From the time it splatters on the ground to when it is extracted by a pump can take thousands of years. Working its way patiently and methodically, it moves a few feet a year.

I knew that ranchers in our groundwater conservation district had reported drops of up to three hundred feet in their static water levels where fracking had been intense. Submersible pumps kept pumping even when the water dropped below the intake. They overheated, burned through the casing, and dropped to the bottom of the well.

There was a lot more to learn. Water extraction isn't even the worst of it: injection wells are. The oilmen blend a carcinogenic cocktail of diesel, water, various herbicides, benzene, toluene, xylene, and other chemicals, then inject it at over 20,000 pounds per square inch. They then pump down the borehole with sand to blast open the fissures that hold what they're after. Once the frack is complete, the fun really starts: flowback. In our particular geologic location we only get roughly 20 percent of the water back. The remainder is lost into God knows where, coursing through strata no one even knows about, never to be seen again. You could say it's lost to the ages, unless of course we have a seismic event, which could send those carcinogens seeping elsewhere, possibly up an abandoned well.

The deck was stacked from the start. At my first official meeting of the water board I found out about an incident that had taken place on April 11, 2011. The board's general manager had gotten a call from Larry Mogford, a local rancher. He had just come in from the pasture and had seen something that was as revolting as it was terrifying: an abandoned oil well oozed a blackish and brownish sludge the consistency of month-old chocolate pudding.

It's called a breakout. Injected frack water from a disposal well that was just over a quarter of a mile away had migrated to Larry's abandoned well, and the pressure from the injection well had pushed the fluid up the casing and onto the ground. Had Larry not discovered the breakout, the toxic sludge, which contained known carcinogens, would have percolated down into the aquifer. He took pictures to prove it. The sight of those photographs of toxic sludge on bare, drought-stricken earth will be with me till the day I die. If that breakout had not been discovered, it would be the end of the story for Dimmit County and the Carrizo aquifer.

What became known as the Mogford Breakout prompted the board to establish a policy of protesting each and every disposal well that was seeking a permit to drill in our district. The visual impact of those photographs of the Mogford was still with me when our general manager said the name Glen Rose. A disposal well was destined for this formation. As a young boy I remembered my father telling me about the Glen Rose and how it held a massive amount of fresh water at extreme depths and temperatures. How, I thought, could we as a water board allow the disposal of carcinogenic waste into a formation that can contain freshwater?

The power of remembrance infused me with a passion that had been smoldering since childhood, all ignited with those two words. It was my trigger, and it was a hair trigger at that. I decided to appear before the Railroad Commission examiners to oppose the company that wanted to inject into an aquifer that in some areas of the district could be a source of drinking water.

On the day of that protest hearing before the Railroad Commission I felt like a man who had little more than truth on his side. I had no lawyer, no hydrologist, and no petroleum engineer at the table with me that day. The other table had two of each along with maps, engineering reports, and seismic logs of corresponding wells. What I had was history. So when my turn to speak came I told the story of a little boy going with his father to see the drilling rig that was boring down more than 13,000 feet into the Sligo formation. It turned out to be a dry hole, but on its way down it encountered the Glen Rose horizon, an oil and gas strata that in certain areas can contain vast quantities of water. The driller took my father into his trailer and handed him a Mason jar filled to the brim with clear, clean water. "It comes out at over 200 degrees, but if you let it cool it's good enough to drink, and there's a lot of it."

This childhood remembrance came tumbling from my lips with a force as powerful to me as it was alien to my nature. I felt like Patrick Henry standing before these learned lawyers and scientists. To my amazement, a compromise was reached with the opposition, who agreed to dispose of the frack water in another formation and leave the Glen Rose just where it has been for the past 7 million years.

Naturally, hubris took its toll. When I tried the Patrick Henry speech again on my next visit before the Railroad Commission examiners in Austin, the opposition pummeled me with maps, pressure gradients, and all manner of scientific and engineering statistics. After leaving the Railroad Commission office, I was so dumbfounded and distraught that it took me thirty minutes to find where I had parked my truck. In the vernacular of the West I had just been given a whuppin'. I limped back down I-35 and reported to my water district. Humiliation is perhaps one of the most effective and powerful tools in interpersonal relations. If native peoples wanted to influence and correct a child, they preferred humiliating to beating. But that gate swings both ways. Humiliation is a powerful driver of revenge, smoldering like a red-hot ember just below the surface.

We got a water lawyer, and we managed to negotiate and cajole injection well companies into resolutions that were much more balanced and favorable for us. Today it remains our single most critical issue; if extraction is moderated, we might recover our recharge and the depletion our aquifer has suffered. But a breakout and contamination of the aquifer from an injection well means game over.

◆ ◆ ◆

Benjamin Franklin proffered a wise warning during the turbulent days of the American Revolution that speaks to our continuing predicament in Dimmit County. When describing the necessity of a unified colonial front against the Crown he said, "We can hang together or we can hang separately." I believe our natural resources are what should unify us as landowners against Big Oil. I asked myself, "Am I ready to become *fully* involved with the community?" The answer was easy. "I'd better be."

When I dressed myself up and strode into the office of the Democratic Party headquarters in Austin on December 10, 2013, I had one thing in mind. I thought it was time to take charge, to bring law and order to the oilfield. The only way to do that was to run for a seat on the Railroad Commission. To put it mildly, I was delusional—but sunlight, the greatest cleansing agent of them all, sent a fireball of light in my direction.

I sat down with the executive director, Will Hailer. I told him what I was all about and what I wanted to do and why. Will is a big guy with a voice that could peel paint. I tagged him with the moniker Jupiter. I liked him immediately. He was from Minnesota and had been in Texas for less than a year. For the next thirty minutes I did a fairly good job of selling him on how great a candidate I would be. I recited facts and figures about the oil and gas industry and how fracking had changed the face of Texas. He looked up at me and sheepishly said, "Hugh, I understood about one-quarter of what you've been saying, so there's no doubt in my mind that you would be a qualified candidate." Wow, I thought. "But we already have someone in mind for a candidate. However, we really do need someone to run for Agriculture Commissioner."

The thought intrigued me. Here was an office that governed and regulated a subject I knew. I could speak about my environmental concerns and at the same time make positive change. I wanted to talk about water issues, school lunch nutrition, and raising food organically to be provided to Texas schoolchildren. I had a lot of ideas.

So they convinced me to run for Agriculture Commissioner. It wasn't all that difficult, and I justified it this way: I would get to speak about the same issues as being a Railroad Commissioner, mainly water and the drought; I would get to feel like I was being heard; and I would get to do what I really wanted to do in the first place—stir the pot and bring the discussion of the degradation of the environment to a boil. I signed on.

Have you ever caught what seemed to be a touchdown pass or hit what looked like an inside-the-park home run, only to be tagged out at third or tripped up on the ten-yard line? That feeling is not too different from being a first-timer in a political race. I launched into my more ambitious political period with all the zeal and determination of a sidelined little boy, a benchwarmer who sat watching the game yelling, "Put me in, coach!" The problem was that I was listening to the cheerleaders instead of looking downfield at my opponents. And cheerleaders make you feel invulnerable, right up to the moment when you get your bell rung.

My advanced political education began simply enough. On December 9, 2013, I drove to Austin, strolled into the headquarters of the Texas Democratic Party, and filed as a statewide candidate in the Democratic primary for Commissioner of Agriculture in the state of Texas. My hat was off my head and into the ring, but my head would never be the same.

My own campaign was not an act of bravery. It was an act of an individual who did not know what the hell he was getting into. In fact, a good friend and fellow rancher asked me when I filed, "Hugh, have you gone crazy?" In a way I had. I bought a ticket for a ride at the carnival. Not too different from the tunnel of love, where equal parts terror and expectation of adoration wrestle with each other until the light of day reveals who is by your side. In my case the seat next to me was vacant when I opened my eyes at 11:30 p.m. on March 4.

I began because I was angry. I had watched my quiet little corner of Dimmit County be overtaken by greed. When I'd made trips to the capital before, all I got was, "Well, Sunny boy, the law says we can do this, so we're going to." In fact, those words were uttered to one of our water board members when he offered to arrange for the Chesapeake Oil Company to take water that had been used to wash spinach and recycle it for use in fracking. "Thanks anyway," they said, "but we won't do that until someone makes us." Why on earth would you listen to someone who wants you to slow down and do the right thing? Because it costs them up to $3 a barrel to truck the water to a well pad, and around $.50 to $.75 to buy it from the landowner, extract it from the Carrizo Aquifer, and inject it down the well bore.

I thought that if I could get elected to the office that regulated the oil and gas industry, then I would have the last laugh. I would be able to demand that these people slow down and start to behave.

By early fall I had gathered a group of landowners and concerned citizens who felt strongly about the impact of oil and gas on our state. We came up with a name: Texans for Responsible Energy Development, or TRED. Our goal was not to stop the industry but rather to encourage it to become more environmentally responsible.

That was no small task. Because there is little regulation of the extraction of underground water in Texas, oil and gas companies

take all they can get. Our state allows the unprecedented depletion of what the law asserts belongs to one owner but which by
its nature should belong to us all.

I am a rancher, a hunter, and a man who loves his land. I should
know better than to expect anything different from a state that
retains an ancient doctrine holding that water below the earth
is private and water on the surface public. We are the only state
west of the Mississippi still bound by the rule of capture.

Economists have a precise name for this situation: they call it
"the tragedy of the commons." It is the absolute worst possible
tragedy for the land, for the landowner, and for living organ-
isms—in other words, everyone and everything that need water.
The rule of capture panders to the most debased side of our nature, and it runs something like this: If you own the land, you
own the water beneath it. You can pump with impunity until the
well goes dry. Especially in the case of small landholdings, you
are most likely to deplete the water from beneath your neighbor's
land—hence the commons. The tragedy comes when a man with
a conscience shows up and says, "I will not pump an excessive
amount because I would not want my neighbor doing that to
me." That's the basis of what is known as the common courtesy
act, which is unenforceable. So it's too bad, too sad if your neighbor across the fence does not share the same ethical perspective.
His view is probably simple: If it's legal and I don't use it and/
or sell it, then someone else will. And so a precious resource that
should be held in common is headed for catastrophic consumption by an outsider, a corporation with a lease in hand, an entity
now protected like a person.

I did not intend for my run to be anything other than a march to victory—a mission to, as Davy Crockett once said, "be the poor man's friend."

I got a dose of both, and my political run was a most magnificent defeat.

◆ ◆ ◆

Six days after my electoral defeat, I was sitting in my home in San Antonio before a small mesquite fire on a quiet March morning. I listened to the songs of cardinals and mockingbirds as I tried to shut out the whine of leaf blowers and the beeping of a garbage truck backing up. Sipping good strong Italian coffee, I reflected on the race, an episode that caused me to step off the cliff socially, psychically, and emotionally. I did not become a Jehovah's Witness, did not run off with an exotic dancer or run in the Boston Marathon. No, all I did as a fifty-nine-year-old rancher was decide to become a politician and run statewide for Commissioner of Agriculture.

I went from cheering throngs and people asking me for my autograph to sitting by the fire and staring into the flames as I tried to copy down the license plate of the bus that ran me over. I was going to practice my mandolin and start a band with two of the other statewide candidates. We were going to call ourselves the Down-Ballot Boys. Instead, I poked the fire.

So what have I learned about myself and the great state of Texas, at least as far as ambition is concerned? How has the intersection of self-deception, runaway ego, and an uninformed and downright ignorant electorate T-boned my psyche? That remains to be discovered. But what is left of me after this experience is what I'm still pondering.

I started all this because I wanted to be heard. My identity for the past sixteen years has been defined by ranching, by raising bison and keeping honeybees, because those things speak to me. I believe that the world needs real food. I also believe that we need each other, and if we don't have one, then we won't have the other. But in the aftermath of my defeat, even writing about it plunged me into a clinical depression. It was a feeling I had never experienced before, just as I had never experienced speaking into a microphone and having hundreds of people clap their hands in response to what I was saying. They were a gateway drug. First you address a group of ten members of the San Antonio Area Food Council, and before you know it there you are, standing before the AFL-CIO's political gathering and raising hell about greedy Republicans and a water policy patterned after archaic English common law. It was all a seductive and powerful siren song that wooed me closer and closer to the edge, until the cliff began to crumble and I lost my balance.

One thing that became clear was that the system we use to elect individuals to public office was so completely broken that it in no way resembled what was intended by the founders of this country. I was defeated by two individuals, one a comedian and the other—well, let's just say he is a man with a pleasant-sounding name.

A friend of mine once told me, "Texas is very young." We have a propensity to ignore the obvious and accentuate the ignorant. In fact, I almost have come to believe after my experience that we as a people celebrate our political ineptitude.

The Democratic Party, because of its history of tolerance and open-mindedness, was the party I aligned myself with. But like some new kid on the block who wants to be everything to everybody, we as a party spread ourselves too thin. Where are the bosses? We need a captain who can sail a ship, not half-assed, half-baked so-called leaders who pander to those who cry the longest and loudest. But it's not going to be me.

· ELEVEN ·

River Journey

I have a habit of going to water when I find myself overwhelmed. After my defeat, I felt a strong need to go with the flow, literally: to give myself to the Rio Bravo and simply be on it, with it.

So maybe I keep coming back to this river to prove myself in a different way. It has always been a place replete with desire, danger, and the thrill of being at the edge of who I am and who I want to become. When I am humbled, I return to be renewed. Just me and the river.

I arrange a meeting place downstream with my ranch foreman, Freddy Longoria, then put my kayak in just below Eagle Pass, where the Kickapoo Nation operates the Lucky Eagle Casino. On the banks there's a new hotel built to house the gamblers, oilfield workers, slot jockeys, and hardcore gamers who have managed to elevate the tribe from cardboard shacks beneath the bridge at Piedras Negras to late-model SUVs and hand-tooled cowboy boots.

It's 9:30 in the morning, gray, overcast, and cool, perfect in every way save for a ten-knot headwind out of the southeast. And thanks to the bends in the river, the wind will only be of periodic consequence.

Freddy helps me slide the kayak down the bank. A recent mid-October downpour has swelled the river and sped up the current. In no time the signs, smells, and sounds of humankind all but vanish. Red-winged blackbirds send their gurgling friendly greeting to one another as I drift. Stopping my stroke I listen and hear what I have been searching for and missing all these years: silence, beautiful, deep. I am here, soaking up the wonder of what there is so little left of in the world.

It almost feels as if I am the first person to behold this beauty. I revel in each bend, every canebrake, bird, and tree I float past. An osprey, the sometime resident of the river, circles overhead and works the air currents as it gains height and perspective. It is hunting for river carp using sun, shade, and eyes that can see the slightest change in the river's ripple from five hundred yards above. The osprey tucks its wings tight and falls into a stoop that would have sidelined Chuck Yeager. Dropping at well over a hundred miles an hour, it flares a split second before impact, talons spread wide, and thrusts with hungry purpose to impale a river carp. With three powerful beats of its four-foot wingspan the osprey is airborne and climbing skyward, its prey flopping helplessly in a grip no human could match. The fish eagle has it all: grace, purpose, and determination.

The carrizo cane grows thick on the American side. The plant was introduced here by the early Spanish explorers more than three hundred years ago and has now permanently embedded itself. Interestingly, the Mexican side of the river does not have half the trouble with the cane as we do. On their side, livestock grazing is allowed, so the cane never has had the chance to become the monoculture it has become for us. The USDA built a fence on our side in order to protect Texas ranchers from the dreaded and deadly fever ticks that can wipe out livestock. All of these fences are permeable.

The river is often in the news for all the wrong reasons. Raw sewage from colonias and the Mexican border towns of Piedras Negras and Nuevo Laredo have sent *E. coli* numbers through the roof. I expect floating fish and the rotting carcasses of both birds and mammals, but what I see is a world that on the surface is the same as when I was last here in the early 1970s. There is of course more trash and debris from recent floods, but no sign of cartels or *alambristas*—wire jumpers—wanting to cross until I round the bend at the hamlet of El Indio, Texas.

El Indio started out as a post office for the Indio Ranch. It was named not for the logical association with Native Americans but for the sharp bend in the river, which resembles the shape of an East Indian turban. At one point in his life my father made inquiries into a land purchase on the irrigation canal here.

A canal off the river imparts hope and the special joy a man takes in knowing that his salvation lies in only a few turns of the wheel that lets the water flow into the field. As the water tower of the village comes into view, I look across the river onto the Mexican side and see what appear to be two men standing next to a small johnboat casting a seine net into the shallows. I watch with my binoculars as they half-heartedly throw and pull the net with no results. They get back in their boat and start coming in my direction. When they're still three hundred yards off, I begin to put the pieces of the puzzle together. I don't like the odds of two to one and them having the small yet distinct advantage of an electric trolling motor. Friend or foe? It's not always easy to tell.

Pulling close, they scrutinize my every move and gesture. I ask them how far the small dam is where I plan to take out. They laugh. *"Muy lejos."*

Not just far but very far away—not the answer I wanted to hear from potential adversaries. But I soon realize they are not here to do me harm. Their intent is to watch the river for signs of law enforcement while pretending to fish. They give me a free pass and silently wave while putting the little electric motor in gear to cruise upriver. The drug and human trafficking is alive and well on this stretch, with its slow current and shallow gravel crossings that connect our two countries. I am grateful not to be seen as a threat to anyone that might be planning an illicit activity this evening.

I put distance between us, and when I round the next bend an unexpected sight greets me. Live oaks four feet around and thirty-foot-high cedar elms flourish on the Mexican side of the bank, signs of better times and more moisture. Under the oaks I see a flock of wild Rio Grande turkeys. The jakes bob their heads and form a bachelor nucleus off to the side while the hens prom-enade; four big old gobblers bring up the rear. They're shy and skittish. Raising my binoculars, I follow them as they hunt and peck for acorns and insects. They're both beautiful and rare in a land where protein is valued and consumed without hesitation.

The distraction recedes behind me. My faux fishermen have made me worry that I have seriously miscalculated how long this journey will be. I have no real clue as to where I am, and the sun is starting to arc toward the west. Every stroke is tempered by the four and a half hours I have now been on the river. So I dig in and start to move faster in brief spurts of power that give way to fatigue. I have no choice but to press on. The prospect of darkness on this river is not a good one. No moon, no flashlight, and no immigration officers when you need them.

But I just can't bring myself to speed up the pace. With every passing moment I try to strike that balance between the fear-sparking reality of kayaking alone down the most dangerous drug-running corridor in the United States and gazing in awe at a natural world that I thought had all but vanished.

A tangerine-orange one-story Mexican hacienda appears on the opposite bank. This neatly kept stucco home has a red tile roof. Three hundred Spanish goats calmly graze around the house. The bell of the lead nanny rings softly as she walks and nibbles at whatever green she can find. It is the most pastoral of scenes, and one I never expected to see. Who could possibly occupy a home in Mexico on a river where there are dozens of shallow fords that would allow any number of smuggling activities to take place? Bucolic surroundings are an effective mask for a trade that dwarfs goatherding, and that house most likely did not come from selling cabrito. I fight off the urge to raise my field glasses and take a closer look given the very real prospect that what might be staring back is an assault rifle with me in the crosshairs.

Over the years I have had a number of encounters with smugglers. Avoidance is the best strategy. A close friend in the security/bodyguard trade once told me, "As long as you never interfere with another man's income, he will not bring you harm." The reality is that along this river there are only two trades for Mexican citizens: running dope or trying to find work in *el norte*. Both are losing propositions from the enforcement perspective.

Yet we seem to be as addicted to the drug war and to ignorant immigration policy as we are to the drugs and cheap labor we have enjoyed since the Mexican Revolution of 1910.

There is plenty of evidence of the trade. I come upon deflated inner tubes scattered up and down the canebrakes, as well as oversized black plastic garbage bags that serve as suitcases for the trade. The bags hang in the cane and swing slowly with the current.

The scolding squawk of a great kiskadee flycatcher snaps me back from my fearful reverie. There it is, landing on a stalk of carrizo cane, its blazing yellow breast and black crown offset by white eyebrows. The kiskadee is one of the original winter visitors, the genuine snowbird that knows no boundary other than hot and cold. On my right, darting along the Mexican shoreline, a belted kingfisher sends out a bold, piercing cry meant to alarm others as to my presence as an intruder. I take the hint and dig deeper with each stroke. These birds have work to do, and it does not include dealing with a visitor equally enchanted by their presence and aware of their interrupted privacy.

A long straightaway flowing directly into a stiff southeast breeze gives way to an oxbow that curves and opens to what must be the premier crossing site on this stretch of the river. Eroded river rocks sculpted by countless floods lie half-exposed, forming a bridge of stepping-stones from Mexico to Texas. Empty backpacks and discarded duffle bags complete the tableau. But it's the baby doll's head wrapped in plastic that really grabs my attention. I pull the boat to the rocks and get out so I can look around, but a rustling in the cane sends me back into the boat. I push through the rapids on the backside of the crossing, stroking with every ounce of strength I have left.

Out here there is no cell service. Besides a useless phone, all I have is a package of jerky, half a jug of water, and an apple. I'm not prepared for anything except an easy outing. I start to wonder, will I be safe back home at the ranch tonight or fine fodder for the snapping turtles that gaze with black beady eyes from the surface of the river? Fear and wonder continue to accompany me on this journey, taking turns.

◆ ◆ ◆

My experience with immigrants on the ranch has been mixed. Back in spring 2005 three members of a gang attempted to break into my home. I stopped them, and while I feared for my life during that one threatening and particularly unpleasant encounter, I wasn't harmed. They were members of the notorious Salvadoran street gang known as Mara Salvatrucha, or MS-13. They have a well-deserved reputation of absolute evil. Their hallmark is torture, intimidating innocents into joining them or else. They are ruthless and gruesome beyond belief. There they stood on the other side of my wooden gate, three rail-thin young men dressed entirely in black, shaved heads and a swath of tattoos, symbols that covered their arms and ran all the way up their necks. When they realized I was armed, they left, but not before hurling insults and a dead rattlesnake in my direction. The Border Patrol caught them three days later, after they had stolen a four-wheeler from a rancher in Frio County.

The next incident occurred early one morning at daybreak.

I had just left my bedroom to walk across the courtyard to the cookhouse. As I approached, the door to the kitchen flew open and a tall, well-dressed young man vaulted off the front porch with a stack of tortillas in one hand and his pink Minnie Mouse backpack in the other. As he vaulted like Jesse Owens over the fence, he turned his head and in perfect English shouted, "Excuse me, sir," then disappeared into the brush. There is nothing like a polite thief to give one perspective on the plight of those less fortunate.

We have had our tragedies. Once we found a dead man, most likely murdered by a companion, his pockets turned inside out and the side of his head crushed by a rock. Then there was the immigrant who walked across the ranch on crutches. We followed the distinctive pugmarks left by only one tennis shoe on the trail. He survived his journey but was turned over to the Border Patrol. Next was the man who was abandoned by his group and left to fend for himself in 110-degree heat with no water. I found him as he was becoming delirious from heatstroke, standing at the end of my driveway and waving his arms madly. He screamed unintelligibly. I lifted him into the back of my pickup and drove to the closest water trough. After five minutes in the water he revived, then asked politely to have me call the Border Patrol so he could turn himself in. He had had enough.

So what has all this meant? Compassion and fear are working on me concurrently. I seem to be more influenced by what might happen than by what actually does happen. But it was another incident that completely rearranged my attitude toward immigration. It was a cold and rainy January night last year, pitch dark—moonless, close to midnight. We were all in bed; the doors were locked and the lights turned out. Then there was a knock at the door.

Now, the one hard and fast rule in the brush country is that you never, ever come up unannounced to a person's home after dark. If you do, you will most likely be shot.

When I parted the blinds I was hoping to see my foreman, Freddy. What I saw was a middle-aged man dressed in dark clothing, sopping wet from the rain and favoring one leg. I opened the door, leveled my shotgun at him, and asked him in Spanish what he wanted. He told me he wanted water, food, and some pliers so he could pull the tasajillo thorns from his swollen knee. I looked in his eyes and then lowered my shotgun. When I did so, he fell to the ground and started sobbing uncontrollably, for at that moment he realized I wasn't going to kill him.

After giving him a bowl of chili and many glasses of water, I drove him to the front gate and showed him the road to Carrizo Springs. By doing so, I broke the law.

We fear what we don't understand. The adage is never more telling than in the current immigration debacle. What we need to do now as Americans is understand why refugees are risking their lives to get to this country. We all have fear, for fear keeps you alive. But unwarranted fear, fear that is groundless and used to instill even greater fear, is what I fear most. There's my fear of dangerous criminals lurking behind the tank dam, then there's the view from the other side: the desperate fear of a man who came to my door for help and thought I was going to kill him. I'm sure that's what he had been told. I can't even imagine the fear a child must experience as he traverses Mexico alone or at the mercy of those who might harm him. What we need to do is put fear in its place. If knowledge is power, then fear will recede and take a back seat behind love. Thank God I looked in that man's eyes.

◆ ◆ ◆

I pick up my paddle and get back to the reality of not exactly knowing where I am other than downstream from where I started. An all-too-familiar drone begins to pound in the distance. I may have come to the river for respite, but it all catches up with me: the sound of high-volume pumps fueled by diesel engines designed to move massive amounts of river water. It's one thing to know the facts of fracking, another thing to be paddling through the same water that now keeps the whole industry going. Suddenly drug runners and cartels seem like schoolyard bullies compared to the gangsters who are hijacking the water from two nations to satisfy the needs of an industry that is desperately seeking to postpone the inevitable.

I had heard that the owners of Anadarko, a new and major player in the Eagle Ford, had chosen to extract water from the Rio Grande, one of the ten most endangered rivers in the world. By pumping it from the river and transporting it through a twenty-four-inch pipeline to their drill pads and frack ponds, they accomplish two critical steps. First, they no longer draw down the Carrizo aquifer, which has already dropped dramatically since the fracking boom began. Second, they are able to fill their ponds quickly, without having to wait for low-volume submersible pumps to amass the 5 million gallons of water needed per frack. A growing body of evidence points to interdependence and the obvious communication of water between the Carrizo aquifer and the Rio Grande. The river is already overcommitted for both agricultural irrigation and municipal drinking water, and now fracking. If this keeps up, it's only a matter of time and drought before it is reduced to a trickle.

All of this goes through my mind as I paddle to the rhythm of the droning diesel engine running the pumps in the distance. This particular energy play is expected to last from ten to thirty years depending on the price of oil and natural gas. Ten thousand more wells at 5 million gallons per frack, and you have a number that short-circuits any calculator. The new paradigm for rainfall is twelve to sixteen inches a year, down from the historic average of twenty-one. The real bottom line is this: without a minimum of sixteen inches a year, there is no recharge. It's gambling on a gigantic scale with incalculable odds.

My destination is a weir dam at the old Paso de Francia crossing. It was first used by Spaniards, who named it for Louis Juchereau de Saint-Denis, a Frenchman intent on establishing trade relations with the northernmost outpost of New Spain. Saint-Denis was an entrepreneur and opportunist of the first order, what my father used to call a "light traveler." Suspected of being a spy, he was immediately arrested and placed in protective custody by the Spanish authorities. Spain was getting rich using Indian slave labor to mine silver in Zacatecas, and Louis probably just wanted in on the action.

Ill will between warring nations initially trumped the smooth charm of the Frenchman. After a brief spell at the Presidio, he was led him off in chains to Mexico City by the commander, Captain Diego Ramón. Saint-Denis languished in captivity until he managed to launch a public relations campaign using the talents of his manservant, who convinced Mexico City's movers and shakers that the Frenchman had no designs for anything other than a trade route that would benefit both countries. Once freed, Saint-Denis traveled back north to the Presidio to attend to a pursuit that suited him more than international trade.

Before his ignominious departure for prison, Louis had managed to ingratiate himself with Captain Ramón's gorgeous granddaughter, Manuela, who helped secure her lover's release. The couple was united at the Mission San Bernardo, a half mile from where I'm floating. Saint-Denis had come full circle: the man who had once marched him off to prison was now walking down the aisle of the church to give him his granddaughter's hand in marriage. I could learn a lot from this man on the power of patience, persistence, and charming one's adversaries.

As I approach the little weir dam, Freddy and my neighbor's ranch foreman are standing at the bank at our prearranged meeting place. In the fading light I can just discern the outline of a shadowy gathering across the river in Mexico. Young men who will not meet my gaze stand behind small mesquites, thinking that somehow they will not be noticed. They shift uneasily from one leg to the next. They watch with a mixture of trepidation and curiosity as a pale, middle-aged white guy struggles to dock his kayak on the bank.

There is almost no light now, and I'm very happy to see my friends. After more than ten hours of paddling, I'm close to exhaustion. My fatigue is magnified by what I've seen on the river—all that pumping, all that desperation. As I stand up at the edge of the weir dam, the unexpected rush of blood to my head causes me to lose my balance. Scott and Freddy grab me just before my head hits the concrete of the weir.

What a finale. Surviving water moccasins, drug smugglers, Big Oil, and rapids, only to be felled by my own head.

· TWELVE ·

Legacies

The use of massive hydraulic fracturing in Texas has risen steadily since around 2008, and by 2014 Dimmit County was subjected to the heaviest frack-water mining in America. In effect, we were being drained. But an even harsher dilemma arose: what to do with the contaminated wastewater from the frack.

The depletion of our aquifer is the one side effect that most people understand, but according to groundwater hydrologist Ronald Green, it's the injection wells that are truly the "ticking time bombs of the Eagle Ford," along with every other shale formation from Pennsylvania to North Dakota and beyond. As long as turning on the faucet still causes a stream of water to come forth, then all is well in the water world. Unless, of course, you drop a line down the borehole of your well and you see that it now rests at 179 feet when last month it was at 160.

I recently invited my rancher neighbor JR over to the Samaniego house on my ranch, the house where once the cowboy Pablo Samaniego raised a family of six in a home of less than nine hundred square feet. JR and I had had some times together involving rum and wild bets. One moonlit night, after working our way to an altered state thanks to copious amounts of Bacardi, we came up on a jackrabbit frozen in the headlights. JR bet me he could catch that rabbit with his bare hands. Thinking I was in for some easy money, I sat in the truck while he went for the unfortunate victim. Before the rabbit knew what happened, JR had him by the hind legs. He held him up in the headlights for me to see. He held out his hand in a "pay me" gesture. I never underestimated him again.

At the age of eighty he's still going strong. I wanted to get JR's take on how our little corner of Dimmit County has changed since fracking came to town. JR has done it all down here: rancher, oilman, rock quarry operator, general do-it-all and know-it-all. The thing is, he really *does* know what he's talking about.

When we sat down to visit I mentioned his spry appearance, and he replied, "Heaven can wait as long as I'm atop the Carrizo Sands."

"First of all," he said, "it's the surface area that's required to drill these wells. Where the conventional vertical well might need a half-acre for drilling and production, the pad site for unconventional would require ten times that. You're talking about putting five to six wells per pad site, with forty-acre spacing between wells. The result is a fragmentation of the landscape, with long-term effects including soil loss, increased water runoff, and the disappearance of over 2 million tons of topsoil a year from wind erosion."

I did the math. If carried out to its maximum capacity, it won't take long for the ranch to become one giant caliche parking lot. And with an average of more than fifteen hundred large eighteen-wheelers needed to drill and produce each well, you've got to have an area where trucks can safely maneuver.

As if damage to the land and the water weren't enough, JR

told me that what's happening to the air hasn't even registered as a problem. He's adamantly opposed to the pernicious and wasteful procedure of flaring gas. The oilman wants the oil, and in this economic climate the natural gas produced is primarily a headache for the producer and a nightmare for the surface owner. Vented and/or flared from a flare stack at the pad site, a twenty-foot orange ball of fire sets off an eerie glow in the night sky. The Eagle Ford flares are so numerous that they can clearly be seen by satellite from outer space. But unless the landowner has the foresight to require that a pipeline be built prior to the drilling of the well, there's going to be flaring. That pipeline is not an unreasonable requirement to include in a surface use agreement because the chances of drilling a dry hole in unconventional drilling are virtually nonexistent. Once the turn has been made and you're in your target formation, it becomes a mining operation. But without a pipeline, the gas has to be dealt with some other way. The solution, according to JR, would be to either sell it, which would mean laying a pipeline before the well is completed, or reinject it. Both of these options would require either prior planning to the drilling of the well or a capital expenditure to eliminate the need for venting or flaring. Royalty owners are getting hosed by allowing their gas to be flared. It's a resource that took nature 500 million years to make, and up it goes in smoke.

The flame that shoots in spurts twenty feet into the air is preferable to smoking flares that emit carcinogenic, volatile organic compounds. They are known as the BTEX group: benzene, toluene, ethylene, and xylene waft over much of the Eagle Ford. If you're within one mile and downwind, get ready for either an asthma attack or a cranium that feels like it has been poleaxed.

JR and I needed more than a drink to escape this mess. Of all the unintended consequences of unconventional drilling and production, the fouling of the air around us has got to be one of the worst. I have struggled with why that is and why in particular I have had trouble writing about this subject. People, myself included, have tended to take the air we breathe for granted. But this was in my face, literally.

The escape of methane, which is both odorless and colorless, is rampant in the Eagle Ford. You can't see it unless you have an infrared camera. This "greenhouse gas" is a serious danger to the environment because in its raw state it's eighty-four times more potent than carbon dioxide. Methane, while harmful, may be the least of your worries when it comes to other compounds expelled from production facilities. You have no idea that the air you are breathing is harming your health until it's too late.

◆ ◆ ◆

My ranch foreman, Freddy Longoria, was riding his four-wheeler down a ranch road located next to BlackBrush Oil and Gas Company's tank batteries and compressor stations. It was a chilly morning, and he was checking a deer feeder and a bow blind where his oldest son was to hunt that afternoon. Unbeknownst to him, the production unit he rode past was wafting out a steady flow of either hydrogen sulfide or the BTEX compounds. The emissions were coming from flare stacks or "thief hatches," the aptly named port at the top of a tank battery. Because of an atmospheric anomaly known as an inversion layer, where warm air at a higher altitude traps cold air on the surface, Freddy rode headlong into an invisible toxic cloud. As he passed the production unit his eyes began to water and sting; by the time he reached the Number 8 windmill, a quarter of a mile away, his eyes became inflamed and his vision blurred. He managed to keep the four-wheeler on the road and crept back to his home, where he washed his eyes with saline solution. His eyesight returned, but for more than forty-five minutes he was blind.

I immediately called the CEO of BlackBrush and told him what had happened to Freddy. His first response: "I don't know what you're talking about. I've never heard of such a thing." My response was to call the TCEQ and begin the long, arduous, and frustrating task of trying to get the agency to wake up and realize that people's health is being compromised because oil companies deny their culpability. They seem to feel they have a duty only to themselves and their shareholders, that that's where their responsibility ends. What is the bottom line for life?

Litigation by the rancher is usually the last resort. As penance, the violator performs environmental acts of contrition that are somehow supposed to assuage the degradation. If only that were true. The reality is that the state's two regulatory bodies that oversee oil and gas, the Railroad Commission and the TCEQ, are little more than cheerleaders for the oil and gas industry. They do not fine companies in a way that requires them to fix their broken equipment. The laws are there to protect industry, not the people.

I had a new cause. As a means of attempting to arrest or at least slow down the deleterious effects of unwanted emissions from gas compressors, flare stacks, and oil storage tanks, I offered the use of the ranch to a group of researchers at Texas A&M in March 2015. For nine months they measured and analyzed the emission characterization of an aging production facility at the north end of the ranch. They have a device called an Open Path Fourier Transform Infrared Spectrometer, or OPFTIS. I immediately christened this machine "Opie" after the young character on *The Andy Griffith Show*, a boy with an insatiable desire to know the how and why of everything that surrounded him. OPFTIS was developed by the Department of Defense for use on the battlefield during chemical warfare. Opie found what it was looking for.

The study combined a thermal imaging camera and an integrated weather station, and together they measured 262 separate chemical compounds. Seventy-three of those compounds registered as being at or above levels dangerous to human and animal health. Everything from benzene to xylene to carbon monoxide was spiking across Opie's graph. It may not have had an odor, but it left an undeniable trail. Rusty, antiquated equipment whose fittings and seals had served their purpose decades ago had been venting untold amounts of noxious gases into the air. And I had proof, on paper.

Armed with this new information, I sat down with the operator at his production facility. This "facility" was a rusty old tin building that smelled of grease, dust, and oil. In that hot, un-air-conditioned space with peeling linoleum on the floor and a broken microwave teetering on a folding table, we sat across from each other and read the study's findings. The facts were there in black and white and could not be ignored. But in Texas, older production facilities are "grandfathered" and can emit just about anything they want to.

I got lucky. The oilman understood. Perhaps he put himself in my shoes and thought how he would react if this were happening to him. The company repaired the equipment immediately, and the emissions were curtailed.

For now, anyway. In the meantime, the fates intervened. Oil dropped to $40 a barrel. Oil prices, as it turns out, are about as predictable as rain.

•   •   •

When the Eagle Ford—which has already become the Eagle Edsel—first arrived, there was great celebration and fanfare as to how fantastic and magnificent the technology was that brought this about. Finally, with the power of sand, water, diesel horse-power, and carcinogenic chemicals, we thought we could shoot the Saudi Arabians the bird. But no matter how impressive the production figures were from an extraction perspective, they meant nothing when it came to the petroleum-fueled geopolitical dogfight we found ourselves in.

At the height of the boom in 2011, I walked into a convenience store in Cotulla to get a bottle of Topo Chico, the Mexican Per-rier. After placing my drink on the counter and paying for it, I asked the tattooed attendant behind the desk if she would please pop the cap for me. There were half a dozen roughnecks lined up behind me, and this local boy in front of them holding up the line did not amuse them. For a few seconds she rummaged around, but when she couldn't find an opener, she grabbed the base of the bottle and cracked the neck down on the linoleum countertop with a solid *whack*. Two and a half inches of pea-green linoleum went flying across the room as the Topo Chico liberated itself all over me. Without even looking up she handed me the bottle and said, "Next."

At that moment oil was somewhere around $90 a barrel. Motel rooms in the Eagle Ford, if you could get one, were $300 a night, sometimes more. The price of a bean and cheese taco went from $.99 to $2.49, and some roughnecks released a YouTube video that showed mysterious green and orange flashing lights zooming down to their well pad, hovering for a moment, and then disappearing into the dark night. The departure was accompanied by the eerie glow of gas flares illuminating their exit back to outer space.

By February 2016 the rubber band snapped. Oil is still holding at $45 a barrel as I write, and below $75 a barrel it doesn't pay to produce it.

These days you can find a parking place at the Pizza Hut. There are hundreds of empty hotel rooms lining the highway into Carrizo Springs. Bright multicolored banners flap in the breeze in their empty parking lots. "Now open! Reduced rates!" Not a taker in sight. Even the makeshift "Man Camps"—a euphemism for a microwave, a bed, and the Playboy Channel—are empty or have been hauled away. Restaurants where you had to sometimes wait for an hour during the boom now have employees standing around looking at their smartphones. Truckers who made more than $100,000 a year are back to eight bucks an hour at the corner gas pump. It was fun while it lasted.

Silent memorials to the departure of big oil are everywhere: empty doublewides with broken windows, a Laundromat whose sole occupant is an undocumented immigrant standing in the shadows of the Coke machine. Then there's the sports bar down by the Walmart whose owners were so anxious to get up and running that they opted for an inflatable roof so they didn't have to wait for a construction crew that they knew would be unavailable.

There are also more permanent, more solemn reminders of how fast and furious this blitzkrieg of extraction happened. Just east of our property, at the intersection of the two farm-to-market roads that lead to the ranch, stands a large cross made of plastic flowers, what the Mexicans call an *ofrenda* to the life and spirit of a loved one who has passed on. On January 16, 2015, in a hurry to deliver his load of produced frack fluid to an injection well down the road, this particular young man drove right through the intersection in broad daylight at more than seventy miles an hour. His brother bore witness to the crash from the eighteen-wheeler he was driving just behind him. The fiery collision killed the driver and four others, men from Laredo who were returning home from an all-night shift on a rig. The ingredients of this tragedy were speed, sleep deprivation, and a tanker full of highly volatile condensate oil. When the body of twenty-two-year-old Eduardo Peña was claimed, his parents reached inside his jeans pockets and found what he always told them would be there: two one-dollar bills, one in each pocket, one for each parent.

Now we're left with a lower water table and tainted air. The memory of all that fast money that seems to have disappeared just as quickly as it descended, kind of like those mysterious flying discs that briefly glowed in the night as they assessed the ghost town below. The aliens aboard probably shook their heads and took off again.

◆   ◆   ◆

On March 2, 2016, the rakish and flamboyant Aubrey McClendon, the founder and former CEO of Chesapeake, one of the major Eagle Ford players, decided he had had enough. He was facing indictment charges for price fixing, and his number-one financial backer had just severed all ties with him. The noose was tightening. Taking aim at a solid concrete embankment on a two-lane road outside his hometown of Oklahoma City, he stepped on the gas of his Chevy Tahoe and met that immovable object head-on at more than one hundred miles an hour. The crash and resulting explosion incinerated him. His death speaks volumes about how hubris and missteps by men who were worshipped for their bold and seemingly invulnerable lives could be brought down by the instability that has always been the hallmark of the oil business. He left behind a wife and three children. It was a death that set off ripples that are still deep and wide.

The ranchers are still here. I ponder our fate. Oil prices may be down for now, but they have a habit of going up again.

Droughts, too, come and go. So far this year we have received upward of nineteen inches of rain since January 1. My grandfather would be pleased to see his beautiful, soft-brown tiger-striped steers, a Brahma-Hereford cross, belly deep in little bluestem. The grass is waist high, and the bobwhites and blues have paired off. They escort each other like punch-drunk brides and grooms down the aisle. There are so many grasshoppers and bugs of every description that the quail don't know which ones to chase. Rain has returned. A slow-moving El Niño, the Pacific current that is the real soaker for us, has been pounding Texas. I crossed the Nueces yesterday and it was actually flowing. Soldier's Slough, an offshoot of the river, has gone from a wide, parched ravine with bare ground to a muddy delta that looks more like the Ganges. It's a sight that stops you in your tracks. Every rancher looks smart when it rains.

But like it or not, if you are ranching in Texas, feast or famine is what you have signed up for. Those cycles don't apply just to oil and gas production. As a rancher, you're a lot less dependent on the complexities of the free market than you are on the simple fact of rain. The glory of being a rancher rests on the throne of a mythology that has been nurtured in this state since the first book about Texas was published. And I am just as liable as any other Texan to take on the trappings of what it means to be born and raised here. I've got them all: boots, guns, horses, and, yes, oil wells. But it can be a treacherous thing to live past the end of your myth.

My ex-brother-in-law, a car dealer in Uvalde, gave me his take on this situation not too long ago. He said, "You know, Sunny, people come in here with all that new oil money, and they buy the pickup they have always wanted. Then they start moaning and groaning about how the oil trucks and the disposal wells and the waste pits are ruining this country. I tell them to stop complaining and just take all that money and move somewhere else." It's an old story, one that I find as depressing as it is reasonable and true.

As a teacher, a lifelong student of history, and an inheritor of property that has both problems and potential, I wish I had looked back before moving forward. I had the best of intentions but often missed the mark. Paradoxically, at times the target was too big. I just couldn't see it then. I didn't really start to get perspective until I wrote this book.

One thing that I know for sure is that I have always had "an attitude" about Big Oil. I've been quick to condemn it, whether the oilmen deserved it or not, and I've been suspicious of their actions as they did their business on my land. But my attitude needs to change.

What we all need is a transition—a transition from acrimony to a cooperative spirit fostered by solid science, knowledge, and hope. One of the thinkers and doers who have been my guiding lights is the Austrian scientist, philosopher, and mystic Rudolf Steiner. When I was nine years old my mother hung a framed poster in my room that showed a graphic representation of the Bauhaus school of architecture, which Steiner helped inspire. I used to lie in my bed at night before dark and just stare at that wondrous representation of what the human spirit, unleashed, is capable of. But it was Steiner's revolutionary approach to the integration of man, industry, and the natural world that ignited me. He was someone who was not afraid to say that a higher power was a part of everything.

I once heard fracking described as the last gasp of a dying industry. That seems accurate enough. I just hope that last gasp does not suck all the oxygen out of the room, because we *are* all in this room together. So in order to understand how best to change hearts and minds, I need to remember what speaks to us as a society. We are rational Western thinkers and value the results of what we can see, but we also respond to our instincts, feelings that can defy reason. What we feel has a different logic, beyond reason, but that doesn't make it unreasonable. Our modern life can look like a tangled web because everything is still threaded together. If only we could be more observant, and patient enough to see the results of our actions. If only we had enough awareness to coordinate with what's around us, we could see the threads connecting us. And maybe we'd stop tying knots, and maybe even untangle some of the mess we've made.

I don't have the pulpit of a high office, but I still want to be heard. I look for what moves people, what connects us. Which connections really speak to us? I believe what matters is love, food, water, and last but not least air, all of which have been woefully compromised by unchecked commerce, this chasm of consumption that is swallowing us. Most of us also respond to animals both domestic and wild—and how we treat animals, especially the ones we eat, speaks volumes of how much we respect the world around us. In raising these animals humanely and as naturally as possible, and taking care of our whole environment, we let compassion and wisdom govern our lives.

Gandhi said to do what's in front of you. Home seems like the right place to start.

◆ ◆ ◆

I became a grandfather on July 1, 2013. My grandson, Leo Wallace Clark, came into this world seven weeks ahead of schedule as a three-pound, fourteen-ounce baby boy. Even in the neonatal unit, before he ever opened his eyes, I could tell what he was about: determination. To see that boy suckle his mama you would think he was King Kong.

He was and is a miracle of prayer and medicine, but on that day there were really two births. Looking into the eyes of your grandson is like coming face to face with an inner self you always knew was there. It is a purity of force that gives you a second chance.

I was twenty-five years old when my son Asa was born. He came into this world at a whopping ten pounds three ounces. Dwarfing every other newborn at the Floresville Memorial Hospital, he was the only Anglo child in a nursery packed with squalling babies. He had a full head of the most beautiful strawberry-blond hair I had ever seen and a face that would turn Michelangelo's head for a second look. In hindsight, I can say I panicked. He was perfect, and it was up to me to keep him that way. I did what I could with who I was at the time. I've had to learn a lot since then. I'm still learning.

Many Native American nations have a legend that the people learned how to live with each other by watching the buffalo. Although the cow may lead the herd, the bull is not always paying attention. But when a calf is born, everything changes. A mother becomes a sheltering envelope of love and tolerance, stoically standing still while her newborn thrusts against her udder with a head butt that would send an NFL lineman to the bench. Resignation to this natural order has its reward: health, vigor, and independence. As a rancher, what you want to see is a herd where newborns are close to their mother—but not too close. For the bulls the task is simple: protect, breed, then tell the calves, "Watch what I do and follow my example."

Leo came into this world before he was ready, and I hope that when I leave this world my efforts will evince at least some of the determination that he showed under the oxygen tent. When I first met Sarah, my wife, she took me to her parents' home to meet them. There on the kitchen counter was one of those kitschy little flat-sided polished stones with a saying or inscription that is supposed to make you stop and think, and this one did just that. It was a Russian proverb that read, "Pray to God, but keep rowing to shore."

With this in mind I mustered the determination to set aside time for the river again. And there I got a story I want to pass on.

I planned to do the same run I had done alone, but this time I wanted companionship. My paddling partner was Colin Mc-Donald. He's the intrepid soul who in 2014–15 paddled the Rio Grande from its headwaters near Creede, Colorado, where my father and I went fishing all those years ago, down to Boca Chica, Texas, where the river empties into the Gulf of Mexico. On his journey he carried a small vial of melted snow from its source in the Rockies all the way to the sea. He even walked the approximately three-hundred-mile dry stretch south of El Paso, where there is no water today.

I didn't have a vial of snow to take with me for symbolic delivery, but every so often I need to remind myself of the river, of the source of life, to bring an open heart and mind.

It was a warm June morning and recent rain had enhanced our chances for making good time by swelling the normally placid river to a decent depth and flow. Freddy waved good-bye as we floated away from the bank, the surreal outline of the Kickapoo Lucky Eagle Casino watching over us as we drifted away from the present and into the past.

No matter how many times you've traveled the same river, you never know what it holds for you when you glide into its waters. The river is as calm as you are, or as tempestuous as you might need to be. You stroke and watch the world go past, carried by thoughts and the current of what the dam upstream has spilled the night before. Add to that mix your own emotions and dead-end thoughts that on land found no resolution. On the water, sometimes the logjam breaks. The river roils and churns all these fragments until the mix of muscle, time, and movement converge to put you in your proper place. You find yourself in a boat, a vessel of time and space to hold you and your companion as you head downstream to see what the rest of the world is dealing with. Time dissolves on the river as past and present meet. Carp nearly the size of small submarines roll at the surface, then disappear into the muddy depths. With each stroke I shed the present.

Birds are the ever-present harbingers of hope: the males with resplendent beauty, catching light and moving with purpose; the females in the background, attending to the important business at hand. Boattail grackles, giant belted kingfishers, ospreys, red-winged blackbirds. The "Oh my" call of the giant kiskadee flycatcher, hidden in the mass of cane but audible to all, is every bit as jolting as the sight of its plumage: yellow, black, white. Crested caracaras patrol the sky. This carrion-dependent species knows the one true thing about a river: it is where many living things go to die.

My only fear is my uncertainty about what might be lurking behind a fallen log or hanging in the dim light of carrizo cane undergrowth. It's great camouflage for people hiding from the Border Patrol, and it's the same cane I attempted to carve into a spear when I was a child, laying the palm of my left hand open. I trace that scar every time I feel the need to hurry through life and neglect the living.

The flotsam hasn't changed. Suspended aluminum Tecate cans tied to the cane are markers of propriety, setting boundaries for those who are sanctioned to use this passageway and putting on notice those who would dare to enter. Bobbing plastic detritus is held fast against the bank by the current. Abandoned, deflated inner tubes that once ferried those who were bringing equal parts of fear and unbridled hope barely stay afloat.

As we paddle downstream, we pass a group of old Mexican men on the US side who have chosen this Saturday morning to do a little fishing. Unlike the supposed fishermen I encountered on my solo journey, these are the genuine article. Sipping a mid-morning beer and watching their corks bob in the ripples of the river, they seem entirely content with where they are. In my broken Spanglish I ask if they have caught anything. The elder of the group can't resist showing me: he hoists a twenty-inch catfish. Two of the other men wave. The last one stares at a cork that will not bob.

Sound on the water carries. Behind us, upriver, the steady full-throated and unmistakable drone of an aircraft engine breaks the relative silence of our surroundings. It's a Border Patrol airboat, and it bears down on us. Its engine shreds the placid air and fills it with the unmistakable scent of high-octane gas. We wave, and the two officers on board acknowledge our presence with inspection through their field glasses. They're suited up in Kevlar bulletproof vests with two-inch-thick headphones to protect them from the noise their craft generates. Obviously, they're not intent on sneaking up on any would-be refugees or contraband smugglers. Their boat scatters anything and everything around it. They don't have to shout it—their boat does: "Get out of the way," it screams to every would-be crosser. "Go back!" The boat passes, leaving a churning wake behind.

Ten minutes later it returns, sending songbirds skyward and turtles diving for cover.

Their presence has put fear in me, and I ask Colin to steer a bit closer to the American side. As if on cue, not four feet from where my paddle parts the water, two smiling faces pop up, suspended in life jackets. These junior entrepreneurs had stayed hidden under the canopy of cane that extended out over the edge of the river, and when they heard the airboat depart they came out and up, clutching a large mesh black duffle bag stuffed with waterproof containers, maybe some contraband, maybe just food and water. They're not startled or alarmed by two men in a canoe passing down the river we share.

As the light fades and the wind slackens, the river takes on a gauzy haze that softens her features and lulls us into stopping one last time to listen to the silence that surrounds us. At first we hear only the gentle lapping of the water parted by the bow. In the next second, the unmistakable, haunting, high-pitched cry of a flock of sandhill cranes cuts through. With my field glasses I can make out the oscillating V formation as the cranes float through the strata of gray clouds a mile high. They're headed for Mexico or points south, governed by changing seasons across divided countries with an instinct to find what has made them whole since time began.

◆ ◆ ◆

After the journey, I wanted to remember the river as a source, as my impetus for renewal, as something I want to share with my grandson one day. We are all travelers on the river that connect us, but there are very different reasons people and animals and birds come to water. Next time, if I were to bring something with me like the vial of snow my paddling companion had once delivered to the gulf, what would that be? Not snow, but rain. Rain in a bottle, a message for all time.

Leo is the fifth generation on this ranch. What will be the course of life for him? Will he have clean air and water? Will I someday be able to float the Rio Grande with him by my side? I have been drawn again and again to the river because I feel that it, like Leo, has the ability to give me a chance to begin again. This idea gives me peace when I lie awake in the middle of the night, even helps me make peace, somehow, with the modern world. It's important to remind myself of the real source of life by getting closer to it. I want Leo to glide past verdant live oaks and watch redtails soar and circle above us.

I found a scrap of paper the other day buried deep inside a bathroom drawer, one of those reminders that someone writes down and then tucks away because the message is so valuable that you never want to lose it. The sad thing is that we only read them when we happen to stumble upon them. Here is what it said: "What is important in life is not what you have done, but the quality of your love." Words to live by. This could be the gift of grandfathers, a transition from the material world of who you are and what you do to how well you love. Quality carries the day.

These are the tracks we leave behind and the lessons we impart, along with the love we share or withhold, those unspoken ideas and dreams of who we are and what we are becoming. What do we want, and what do we want to give to this life on earth? A grandchild gives you another chance to be yourself.

My father's saying comes to mind: "There is nothing better for the inside of a man than the outside of a horse."

It's true. I have felt it. I guess it runs in the family. So one day I saddled up my old gray horse, Lombraño, and took grandson Leo for his first ride. I don't know who was more surprised or excited, Leo or me. Lombardo took it all in stride.

It was all Leo could do to contain himself. I could feel the happiness radiate from the core of his two-and-a-half-year-old body. He was in my lap, and I was in heaven. We were all connected, Leo, Lombraño, and me. Every bird, every jackrabbit, every blade of grass and bush that swayed in the soft breeze of morning took on a new life of its own as we slowly plodded past my past, through the present, toward Leo's future.

I may have been only six months old when my grandfather died, but I have felt his presence my whole life. If he were here today, riding with Leo and me, we would first make our way through the catclaw and blackbrush, following a bison trail toward the San Francisco windmill.

Next to the windmill my grandfather would pause and shake his head at the sight of the solid rock pila that once held the water from the mill. It was built before he even bought the ranch, a silent testament to all who pass by, a visual proclamation of the strength and fortitude of these rocks and the men who laid them there. At the moment, it's dry. So much has changed.

Leo and I came up on a rise and saw the bison herd grazing in the distance. They made their way through the tanglehead and bluestem, pausing from time to time to check their surroundings to determine where they were in relation to each other. I would tell my grandfather that a bison needs only three things: "Food, water, and each other."

The Laredo pasture is where I would take my grandfather and grandson next. It is a wide-open expansive piece of ground, with native grasses swaying gently in the breeze. From the Number 15 windmill, the highest point in the pasture, we could see the Sierra del Burro Mountains across the river in Mexico. Their purple outline is etched on the horizon over one hundred miles away. The sight of them always fills me with the quiet comfort that comes from knowing there is still a wilderness out there, even though thin trails of smoke from two power plants in Mexico are also discernible. The burning of low-grade lignite coal from these plants sends fumes all the way to West Texas and beyond, reminding me of how little control we have over what is done to our planet: coal mined in Texas is sent to Mexico to be burned, and the soot then catches the prevailing southeast breeze and is lofted back across the border.

For now, Leo and I sit in the saddle, watching the colors of the sunset change with every passing second. The washes of orange, purple, blue, and red give rise to the last feathery wisps of cirrus uncinus clouds, the ones the old-timers called mares' tails. We take our comfort where we can. Sky is something we still have plenty of.

From its gentle sloping rise, 850 feet above sea level, we can lose ourselves in the sandy ocean of what the Creator has laid before us. Grass sways as far as the eye can see. Redtail hawks soar and circle prey that scurry into armored thickets of allthorn and whitebrush.

The sea change that has taken place in this state since the death of my grandfather is but a blink of the eye in geologic time. Cathedral Rock is far older. Would my grandfather think the destruction was worth the riches? For a man who drilled for a living and made his fortune by working hard and being in the right place at the right time, how would he see the extraction of minerals on this land that he loved? He saw firsthand the long-term irreversible effects of exploration and production. Back in his day, he had his own land, and he wanted to protect it from someone like himself, someone with a cable-tool rig and an open pit for the flow of black gold that came to the surface with every stroke of the pump jack. He came back to what he always loved most.

In the words of Wallace Stegner, "I may not know who I am, but I know where I am from." This is where we are from.

We're still here.

And so are the bison.

EPILOGUE

Now that the unforeseen has revealed itself and the shroud of money has given way to the hard truths of environmental degradation, the image in my review mirror is coming into focus. Like a runaway eighteen-wheeler, the fracking boom appeared out of nowhere, and it will not subside unless the price of oil plummets or our water supply is pumped dry. It has always been about the water, from prehistoric times, when survival kept humans tethered to a river, to the present day, as the oil industry pumps billions of gallons of groundwater to frack formations miles beneath my feet, water that will never again see the light of day, entombed in shale or sandstone, lost to the ages.

As if our air and water issues were not enough, I received word at a recent water board meeting that a sand mine is being proposed southwest of Carrizo, where sand will be washed with millions of gallons of fresh water so that it can be used as proppant in the fracking process. The location is adjacent to the historic Carrizo Creek, whose springs gave birth to the town that Levi English, the Bells, the Tumlinsons, and so many other pioneer families relied on for their livelihood. While those springs have been dry for more than seventy years, mining the strata of sand that recharges what's left of the aquifer is tantamount to dese-crating the memory of those people.

With time and an emphasis on agricultural endeavors that complement an arid environment, this land can recover while nurturing us with its bounty. What it cannot recover from is a drop in the water table that makes pumping uneconomical or the poisoning of that water by an injection well. And that is what we face today—the need to devise a way of life within our environ-mental limitations and a land-use ethos dedicated to the long-term survival of the land and its inhabitants. We have followed the myth, and now we find ourselves at the edge of the cliff. The Roman philosopher Sallustius said it best: "Myths are things that never were but always are."

I have confronted the essence of Western economic ideology—that having more will bring you happiness, unless what fuels you and gives your life meaning is the abundance of the natural world, an abundance that comes from living with it in harmony.

Behind me stands the Number 15 windmill, which has always been known as the High Mill, a silent steel sentry whose faded battleship gray fan blades spin round and round, the rudder ad-justing to a breeze that keeps her headed into the prevailing wind. In the far-off distance lie the mountains of Mexico. My father told me this is where he wants his ashes scattered.

My bison, ambling together through the native bluestem, tan-glehead, and hooded windmill grass, are belly deep in blissful satisfaction with what surrounds them. At sunset they move sin-gle file up the hill to the water before they retire for the night, oblivious to the troubles of the twenty-first century that have plagued our state. A meager three or four gallons a minute dribble into the pila—just enough to water stock and wildlife, to keep this place a destination for birds and bugs. Abundance is built from the bottom up, and water is the source of that wealth.

As the planet warms, as weather events become more extreme, I feel a responsibility to bring to light those factors that have irrefutably contributed to the earth's troubles and to ours, which are one and the same. Climate scientists at NASA recently established a direct correlation between fugitive methane from oil and gas production and a rise in global temperatures. When the price of oil declined in 2014 and drilling slowed, so did the rise in atmospheric temperatures. Conversely, with the recent rebound in oil prices, the drilling and the global temperatures both increased. What is happening here in South Texas is partly responsible for the fix we're in.

The fix is actually in too. Since 2008 Wall Street investors have lent more than $250 billion to oil and gas companies who pursue unconventional shale drilling. These companies have burned through this capital, managing to lose $280 billion more than they have taken in. Why does investor capital continue its unabated flow? Because banks charge a fee on loans that they make; it's called the vig.

Companies may not make a profit, but the fee is all that really matters to the banker. Debt is the lifeblood of the unconventional energy industry, and just as the 2008 housing bubble burst, it's only a matter of time before this scheme implodes, leaving landowners holding the bag. Depleted aquifers, fragmented landscapes, and compromised air are the next generation's unintended legacy, one that no responsible parent would ever want to pass along.

How do you help people see that what they do harms those things that mean so much to us all—the water and clean air that we need to exist? You feed them. The land's hope lies in the concept of a restoration economy, a concept that is gaining momentum throughout the West. Cultivate crops like olive trees, harvest the abundant mesquite bean, and utilize medicinal desert plants such as creosote, lechuguilla, and the edible tunas from the prickly pear.

There is an ethereal satisfaction that comes from eating real food that is raised or grown with love. It transmits life to life, and it requires clean water to do so. Sharing sustenance builds community. I know this from personal experience: raising food and feeding people connects me to others in a way that is truly mysterious. When I was in second grade I took vegetables from the ranch garden back home to San Antonio, put them in my little red wagon, and took off through the neighborhood, hawking my bounty door to door. I was a pintsized peddler who knew every soft touch on the block.

At one time Dimmit County produced and shipped boxcar loads of spinach, carrots, onions, strawberries, and citrus, and we had community. But today the Carrizo aquifer is rapidly being dewatered. The flowing springs the town was named for are a phantom of their former selves, a distant memory surrounded by the steel straws that pierce the bedrock, draining it until there is no more. For the rule of capture, that ancient edict from the Roman Empire, has killed the commons.

People refer to life-changing events as watershed moments for a reason, and unless you want to have your life changed for the worse you better pay attention to your watershed. The two watersheds my land is bound by—the Nueces River and the Rio Grande—are as different as they are instructive. The Nueces is born in the rocky limestone regions to our north, and the Rio Grande, a Colorado river that dies near El Paso only to be born again by the Rio Conchos, is an infusion granted to us from Mexico.

Moving water is a force of nature that has no equal. The sensation of being touched by a river, and its sound and smell, elevate us to a higher plane. The prosaic and the profane are shed here, the accumulated detritus of our troubles magically cast away. *Recuerdos*, memories—this is where they are born.

ACKNOWLEDGMENTS

For their help and encouragement along the way I would like to thank the following:

Marguerite Avery
Scott Field and Evelyn Bailey
Martha Beard
Dr. Donald Clark
My children: Asa, Patrick,
    and Evelyn
My father, Hugh A.
    Fitzsimons Jr.
Robert Flynn

Allan Kownslar
Martin and Heather Kohout
Freddy Longoria
Marshall Miller
Allison Moore
Sarah Nawrocki
My mother, Lolly Negley
Tom Payton
Barbara Ras

CAPTIONS

All photographs appear courtesy of the Fitzsimons family.

Preface. Me at the San Francisco windmill with Ready in 2011, the driest single year on record in Texas. The well, with one of the last wooden tower windmills in the county, was drilled in the late 1920s.

Chapter 1. A member of my grandfather's 1925 Lake Maracaibo Venezuela oil expedition standing at an open pit of petroleum.

Chapter 2. My father on the bottom rung of the windmill and a fellow junior wrangler above him.

Chapter 3. Cathedral Rock, a beacon of sandstone standing the test of time.

Chapter 4. My father in the rodeo parade in downtown San Antonio, 1940 or 1941. I believe his pony was named Cavalcade.

Chapter 5. My first fishing trip with the men. I'm posing with my prize catch on the banks of the Rio Grande in 1964. Just moments later, this catfish knocked me to the ground with a slap of its tail.

Chapter 6. Bill Hudson, my father, Gus White Jr., and Colonel Watson on the polo field.

Chapter 7. My father, at my grandparents' home in Alamo Heights, with his first two whitetail bucks taken by him in Dimmit County.

Chapter 8. My grandfather with the Lake Maracaibo oil executives in 1925.

Chapter 9. Frack pond being filled by my neighbor to the north.

Chapter 10. Ripe cactus tunas.

Chapter 11. Two bison sisters gazing and grazing.

Chapter 12. Third- and fifth-generation Texas ranchers Hugh and Leo astride Tractor.

Hugh Asa Fitzsimons III is a third-generation rancher from Dimmit County, Texas, and a director of the Wintergarden Groundwater Conservation District, which focuses on preserving and protecting groundwater across three counties. He holds a master's degree in history from the University of Texas at San Antonio, and he is the owner of Thunderheart Bison and Native Nectar Guajillo Honey, established in 1998. He is currently expanding into the cultivation of environmentally beneficial restorative plants and crops.